ULTRASONIC BIOINSTRUMENTATION

ULTRASONIC BIOINSTRUMENTATION

Douglas A. Christensen
University of Utah

JOHN WILEY & SONS

> A NOTE TO THE READER
> This book has been electronically reproduced from digital information stored at John Wiley & Sons, Inc. We are pleased that the use of this new technology will enable us to keep works of enduring scholarly value in print as long as there is a reasonable demand for them. The content of this book is identical to previous printings.

Copyright © 1988, by John Wiley & Sons, Inc.

All rights reserved. Published simultaneously in Canada.

Reproduction or translation of any part of
this work beyond that permitted by Sections
107 and 108 of the 1976 United States Copyright
Act without the permission of the copyright
owner is unlawful. Requests for permission
or further information should be addressed to
the Permissions Department, John Wiley & Sons.

Library of Congress Cataloging in Publication Data:

Christensen, Douglas A.
   Ultrasonic bioinstrumentation.

  Bibliography: p.
  Includes index.
   1. Diagnosis, Ultrasonic—Instruments. I. Title.
RC78.7.U4C48   1988b   616.07′543   87-34066
ISBN 0-471-60496-8

To
Laraine
and
Ithaca

Preface

The purpose of this text is to present the physical and engineering principles that underlie the use of ultrasound in medical instruments. The text is intended to be helpful to both users of ultrasound in the clinic as well as to designers of ultrasonic instruments. It is felt that only by understanding the mathematical basis for the propagation of waves, including the development and solution of the wave equation and the study of such phenomena as reflection, impedance, and radiation patterns, can the user of ultrasonic bioinstrumentation knowledgeably apply these concepts in clinical practice. Similarly, a bioengineer who is designing new devices needs to obtain a physical understanding of waves and their propagation behavior in order to exploit these ideas in future instruments. The first chapters in this text develop the fundamentals of ultrasound, and later chapters apply these basics to practical instruments.

The level of the text is appropriate to a senior or first-year graduate course. It is assumed that the student has had some training in partial differential equations and has been exposed to the use of vectors. A student who has also had some experience with wave concepts from an undergraduate electromagnetics or optics course will find the analogies between the propagation and reflection of electromagnetic waves and acoustical waves to be reinforcing and reassuring in the study of ultrasound; however, such background is not required for this text.

It should be noted that no attempt has been made to be comprehensive in the wave equation. For example, nonlinear and second-order terms are neglected, as being of small magnitude; that simplification is made in order to concentrate on the primary principles. Also, there has been no attempt to survey completely all current clinical instruments in use today. Such an effort would quickly be outdated by the rapid proliferation of clinical machines seen recently. Instead, typical examples are given of the most general instrument classes, and it is hoped that the basic principles presented can be readily extrapolated to understand instruments now and of the future.

Douglas A. Christensen

Contents

Useful Physical Constants and Conversion Factors   xiii

1   INTRODUCTION   1

2   THE WAVE EQUATION AND ITS SOLUTIONS   5
2.1   Derivation of the Wave Equation   5
2.2   Solutions to the Wave Equation   10
2.3   The Wave Nature of the Solutions   11
Problems   18

3   IMPEDANCE, POWER, AND REFLECTION   21
3.1   Introduction   21
3.2   Impedance of a Medium   21
3.3   Power Density   24
3.4   Reflection of Waves at Interfaces   26
    3.4.1   Angles of Reflection and Transmission   26
    3.4.2   Magnitudes of Reflected and Transmitted Waves   29
    3.4.3   Powers of Reflected and Transmitted Waves   34
Problems   38

4   ACOUSTICAL PROPERTIES OF BIOLOGICAL TISSUES   43
4.1   Introduction   43
4.2   Survey of Biological Tissues   43
    4.2.1   The Cell   44
    4.2.2   Tissue Types   45
4.3   Attenuation in Tissues   53
4.4   Viscosity Relaxation in Tissues   59
4.5   Values of Acoustic Parameters for Biological Tissue   60
Problems   64

5 TRANSDUCERS, BEAM PATTERNS, AND RESOLUTION 69

- 5.1 Introduction 69
- 5.2 Electrical Excitation of Piezoelectric Transducers 70
  - 5.2.1 Continuous Wave Excitation 72
  - 5.2.2 Pulsed Excitation and Axial Resolution 82
- 5.3 Beam Patterns 89
  - 5.3.1 Near-Field Pattern (On-Axis) of a Circular Transducer 91
  - 5.3.2 Far-Field Pattern of an Ultrasound Transducer 94
- 5.4 Width of Beam in Near Field and Far Field 102
- 5.5 Focusing with Lenses, and Lateral Resolution 104
- 5.6 Linear Arrays 112
- Problems 118

6 DIAGNOSTIC IMAGING CONFIGURATIONS 123

- 6.1 Introduction 123
- 6.2 A-Mode 124
  - 6.2.1 Signal Conditioning 129
  - 6.2.2 Applications of A-Mode 133
- 6.3 B-Mode 135
  - 6.3.1 Compound B-Scanners 140
  - 6.3.2 Applications of B-Mode 141
- 6.4 Real-Time B-Scanners 143
  - 6.4.1 Mechanical Scanners 144
  - 6.4.2 Electronic Scanners 146
- 6.5 M-Mode 158
- 6.6 C-Mode 160
  - 6.6.1 Acoustic Holography and Synthetic Aperture 163
  - 6.6.2 Ultrasonic Computerized Tomography 165
- 6.7 Other Medical Uses of Ultrasound 166
  - 6.7.1 Surgery with Ultrasound 166
  - 6.7.2 Heat Production and Hyperthermia 167
  - 6.7.3 Ultrasonic Microscope 169
- Problems 169

7 DOPPLER AND OTHER ULTRASONIC FLOWMETERS 175

- 7.1 Introduction 175
- 7.2 The Doppler Principle 176
  - 7.2.1 Vector Formulation 179
  - 7.2.2 Examples of Doppler Configurations 181
- 7.3 CW Doppler Flowmeters 185
  - 7.3.1 Audible-Output Instruments 188
  - 7.3.2 Analog-Output Instruments 190
  - 7.3.3 Spectrum-Output Instruments 192
- 7.4 Doppler Imagers 192
  - 7.4.1 CW Transverse Scanners 194
  - 7.4.2 Pulsed Doppler Flowmeters 196
  - 7.4.3 Duplex Scanners 200

7.5 Transit Time Flowmeter 203
7.6 Vortex Flowmeter 207
Problems 208

8 THE SAFETY AND MEASUREMENT OF ULTRASOUND 213

8.1 Introduction 213
8.2 Possible Mechanisms of Damage 214
8.3 Measuring Ultrasound Exposure Levels 215
    8.3.1 Measurement Techniques 218
8.4 Safety Standards 220
Problems 223

BIBLIOGRAPHY AND SUGGESTIONS FOR FURTHER READING 225

INDEX 231

Useful Physical Constants and Conversion Factors

USEFUL PHYSICAL CONSTANTS

Planck's constant: $h = 6.63 \times 10^{-34}$ Js $= 4.14 \times 10^{-15}$ eVs
Speed of light in free space: $c = 3.0 \times 10^8$ m/s
Dielectric constant of free space: $\epsilon_0 = 8.85 \times 10^{-12}$ F/m
Speed of sound in air (at 1 atm and 20°C): $c = 344$ m/s
Speed of sound in water (at 27°C): $c = 1501$ m/s
Specific heat of water (at 37°C): $S = 4.18$ J/g°C $= 0.998$ Cal/g°C

CONVERSION FACTORS

1 liter (L) = 0.001000027 m^3 ≈ 1 × 10^3 cm^3
1 newton (N) = 10^5 dyn
 = 0.10197 kg weight (at gravitational acceleration
  $g = 980.665$ cm/s^2)

Chapter 1

Introduction

The use of ultrasound in the modern-day medical clinic has found a solid niche among the various methods for imaging the body. The reasons for this popularity are many, but they perhaps chiefly derive from the ease and safety associated with its use. Ultrasound is defined as acoustic waves with frequencies above those which can be detected by the ear, from about 20 kHz to several hundred MHz. Medical instrumentation typically uses only the portion of the ultrasound spectrum from 1 MHz to 10 MHz due to the combined needs of good resolution (small wavelength) and good penetrating ability (not too high a frequency). The waves are generated by small acoustic transducers, usually hand-held, that are electrically driven and placed on the surface of the skin. The waves propagate into the tissues of the body where a portion is reflected from the myriad of interfaces between tissue types of different acoustic properties. Some of these interfaces are abrupt, representing major organ boundaries, and some are more gradual.

In one common instrument configuration, the B-mode, the transducer is pulsed so that the reflected waves come as a series of various amplitude echoes whose arrival times after the transmitted pulse represent the depth of the reflecting boundary. In another configuration, known as Doppler velocimetry, the reflecting surfaces are moving (such as red blood cells in a pulsating blood vessel) and the reflected sound waves are shifted in frequency proportionally to the velocity of these scatterers. In the following chapters the principles of these instruments and more will be discussed in detail.

Since the sound waves are generated external to the body and no foreign substances need to be introduced into the body to interact with the waves, ultrasound is considered to be a *noninvasive* technique. This is opposed to other techniques, such as indwelling sensors placed in position surgically, which require piercing the body's outer surface. That there is no need for surgical intervention represents an important advantage of ultrasound. When coupled with the relatively straightforward electronic apparatus required for transmission and display, ultrasound is noted for its ease of use.

Ultrasonic imaging is often compared to radiography (imaging with X-rays) as to the effectiveness and accuracy of the images produced. Two major differences are evident. The first relates to the safety of the procedure. X-Radiation has a well-documented hazard associated with its use, depending upon the dosage required, which means that a judgment must be made as to whether the benefit obtained outweighs the risk. This is especially true for obstetrical and gynecological cases, as X-radiation is most harmful to cells in the dividing phase. Ultrasound, on the other hand, is generally considered safe, even for pregnancy scans, at the low power levels used in routine imaging. (At much higher levels, some cell damage occurs with ultrasound, and more study needs to be done on its long-term hazard at high powers.)

The other major difference between the two techniques relates to the characteristics of the tissue that is actually imaged by each. X-Radiation is attenuated (absorbed and scattered away from a straight-line path) in a power law relationship according to the density of electrons in the tissues encountered along the X-ray's path between the generator and the film, so a traditional X-ray view shows differences in tissue density as differences in exposure levels on the film. Because of this, dense tissue such as bone is delineated nicely from other less dense tissue such as muscle. But unfortunately, there is only a slight difference in X-ray attenuation between most types of soft tissue and therefore a radiographic view will not differentiate very well between various soft tissue types. For example, a soft tumor may be hard to see when it is located internal to a soft tissue organ, unless a radio-opaque contrast dye is invasively injected into the region.

Ultrasonic waves, on the other hand, are easy to launch through soft tissue and liquid, and, as mentioned earlier, they are partially reflected at interfaces between different soft tissue types. Thus, an ultrasound scan may be more sensitive to variations in soft tissue type than a radiograph, which is another reason why it is often used in obstetrical scans. It must be noted, however, that the advantage that ultrasound possesses in soft tissue scans is counterbalanced by the fact that ultrasound will not penetrate bony areas or air spaces readily, which makes it impractical for scanning the lungs (at least at the higher frequencies) and for imaging regions blocked by bone, such as the brain behind skull. These points are discussed in a later chapter.

INTRODUCTION

When one considers the different physical tissue properties that are imaged by ultrasound as opposed to X-radiation, it can be seen that ultrasound and radiography should not be viewed as competitors. Rather, they are complementary procedures, since both may image regions of the body from different viewpoints, being sensitive to different tissue properties. What one technique misses, the other might expose. Bone fractures are best viewed with X-rays; pregnancy scans are best done with ultrasound; suspected abdominal tumors or heart valve abnormalities may call for both techniques. Figure 1.1 summarizes the unique characteristics of each modality.

Summary of Chapters

The first portion of this text develops the mathematics of ultrasonic wave propagation and discusses some of the physical principles involved. In Chapter 2 the lossless wave equation is derived describing the propagation of the important compressional waves through a supporting medium. So-

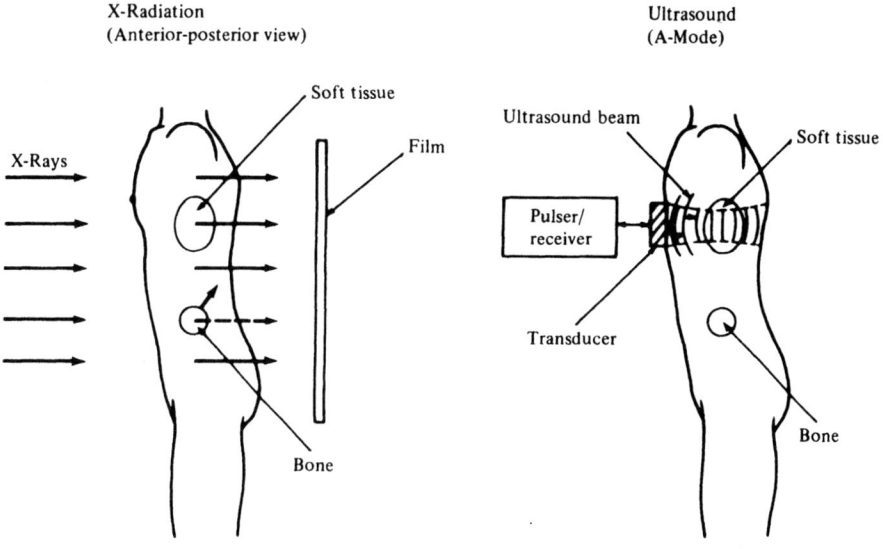

Figure 1.1 A comparison of imaging by X-radiation to ultrasonic imaging.

lutions to this differential equation are presented and are examined for wave behavior, including a discussion of the interaction of phase velocity, wavelength, and frequency. Chapter 3 examines the critical topic of reflection of acoustic waves at boundaries, and equations are derived to predict the amount of reflected and transmitted power at any given tissue interface.

Chapter 4 begins with a brief survey of tissue types found in the body. A quantitative description of the acoustic properties of these various tissue types is presented, allowing the numerical calculation of returned powers from a realistic model of an ultrasonic imager. At this point an improvement in the wave equation is made, adding terms which represent the loss mechanism in biological tissue and examination is then made of the effects of this ultrasonic attenuation in tissue.

Chapter 5 deals with the acoustic transducers that generate sound waves from an electrical signal, and vice versa. Particular attention is paid to their frequency dependence and to the resulting beam patterns. The important concept of spatial resolution (both axial and lateral) is derived here.

The next two chapters deal with the design of practical ultrasonic bioinstruments. Chapter 6 discusses the classification of imaging instruments into general categories depending upon display mode and transducer placement. Examples are given of each of the categories. Nonimaging uses of ultrasound in medicine are also introduced.

Chapter 7 describes the use of ultrasound to measure flowrates in the body. The chapter starts with a derivation of the important Doppler principle, then applies this to practical Doppler blood flowmeter design. Other ultrasonic flowmeters, such as the transit time flowmeter, are also analyzed.

The last chapter covers what is known today regarding the safety of ultrasound, and describes various techniques for measuring its exposure power levels.

The reader should pay particular attention throughout this study to a characteristic of ultrasonic instrument design that is found again and again throughout the entire realm of engineering design, namely, the tradeoff that usually needs to be negotiated between conflicting instrument requirements. A good example of this is the tradeoff between the need for an adequate depth of penetration of the ultrasonic wave into the body and the requirement for good resolution in the resulting image. It will be seen that a compromise in the ultrasound frequency is needed here. Other examples of design optimization will be pointed out as the principles evolve.

Chapter 2

The Wave Equation and Its Solutions

2.1 DERIVATION OF THE WAVE EQUATION

Acoustic (sound) waves are merely the organized vibrations of the molecules or atoms of a medium that is able to support the propagation of these waves. When the frequency of the vibration is above the audible hearing range, the waves are known as ultrasonic radiation. As their frequency is increased, the wavelength of these waves gets progressively smaller, and this small size accounts for some of the unique resolution capabilities of ultrasound when compared to ordinary sound waves.

To develop the acoustic wave equation, consider an incremental volume element of the supporting material as shown in Figure 2.1. Assume that this volume size is small compared to the wavelength of the waves so variations of quantities throughout the volume are slight, but that is large enough to contain many molecules or atoms so the material can be considered to be made up of continuous "particles." The length of the tiny volume along the longitudinal axis is Δz, and the area of the faces perpendicular to the longitudinal axis is A. Since the density of the material is given by ρ, the mass of this volume is given by $m = \rho A \, \Delta z$. Since the volume size is small compared to a wavelength, ρ undergoes only a small change throughout the volume and thus may be approximated here by a single quantity. There is a variation in the pressure in the longitudinal direction in the material, so that the excess pressure above static pressure at any z position is given by $p(z)$. In conjunction with the pressure field and coupled to it, there is a longitudinal velocity field $u(z)$ representing the local motion of the particles of the medium.

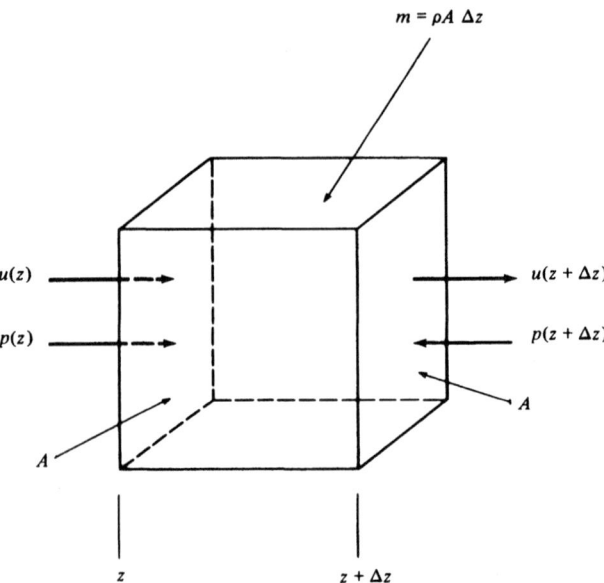

Figure 2.1 Pressure and velocities in an incremental volume of material.

It is important to note that by restricting both the particle velocity and the variation of this velocity to be solely in the longitudinal direction, we are considering here only so-called longitudinal waves, also known as compressional waves, and are ignoring initially any transverse motion which describes transverse, or shear, waves. We do this because compressional waves are the most important kind of ultrasound waves for biological purposes (shear waves damp out quickly in all tissue except bone), and an understanding of most of the wave principles can be obtained by analyzing the simpler compressional waves. Figure 2.2 shows that for compressional waves, the motion of the particles is in the same direction as the propagation of the wave, leading to adjacent areas of compression and rarefaction in the material (hence the name compressional waves), while for shear waves the particle motion is perpendicular to the wave propagation direction. Also, since the only variation considered for our compressional waves is in the z direction, the wave equation that results is known as a *one-dimensional* wave equation.

Newton's force equation can be applied to the volume element. In partial differential form using Eulerian coordinates,*

---

* Eulerian coordinates are assumed fixed in space such that the material particles flow past the coordinate positions, as opposed to Lagrangian coordinates, which follow the displacement of a specific particle.

2.1 DERIVATION OF THE WAVE EQUATION

$$F = m\frac{du}{dt} = m\left(\frac{\partial u}{\partial t} + \frac{\partial u}{\partial z} \cdot \frac{\partial z}{\partial t}\right) \quad (2.1)$$

Since $\partial z/\partial t = u$, Equation (2.1) becomes

$$F = m\left(\frac{\partial u}{\partial t} + u\frac{\partial u}{\partial z}\right) \quad (2.2)$$

Since pressure is force per unit area, the net force on the volume is

$$F = [p(z) - p(z + \Delta z)]A \quad (2.3)$$

Substituting Equation (2.3) and the expression for mass into Equation (2.2) gives

$$\frac{p(z) - p(z + \Delta z)}{\Delta z} = \rho\left(\frac{\partial u}{\partial t} + u\frac{\partial u}{\partial z}\right) \quad (2.4)$$

Taking the limit as $\Delta z \to 0$ gives the differential equation

$$-\frac{\partial p}{\partial z} = \rho\left(\frac{\partial u}{\partial t} + u\frac{\partial u}{\partial z}\right) \quad (2.5)$$

Now it is assumed (with good justification for the power levels normally encountered in imaging instruments; see Problem 3.5) that the *variation* in density ρ of the material due to the action of the waves is a very small

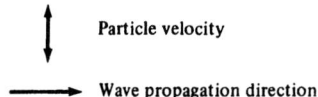

Figure 2.2 Directions of particle velocity and wave propagation for two types of ultrasonic waves, compressional and shear. Compressional waves have much less loss than shear waves in soft tissue and are the only type analyzed in the text.

percentage of the average unperturbed density. That is, if the density is written as the sum of two parts,

$$\rho = \rho_0 + \rho_1$$

where

ρ_0 = average material density (a constant)
ρ_1 = time-varying change in density

then it may be assumed that $\rho_1 \ll \rho_0$. Furthermore, the quantities p, ρ_1, u, $\partial u/\partial z$, and $\partial u/\partial t$ all represent small variations, so that whenever two of these variables appear together as a product in an equation, this product is second order and can be neglected compared to other first-order terms in which there is only one of these variables; Problem 3.5 justifies these approximations for typical imaging power levels. Using this reasoning, Equation (2.5) can be reduced to

$$\frac{\partial p}{\partial z} + \rho_0 \frac{\partial u}{\partial t} = 0 \qquad (2.6)$$

Equation (2.6) gives a time–space relationship (using only the unperturbed density ρ_0) between the pressure and velocity in the medium, and was developed from the force equation. Another equation relating these two quantities can be obtained by invoking the conservation of mass law. This says that the net mass leaving the incremental volume through its sides must be accompanied by a decrease of mass inside the volume. The net mass leaving the volume per unit time is given by $A[\rho(z + \Delta z) u(z + \Delta z) - \rho(z) u(z)]$, and this must be matched by a rate of change of mass given by $-A \Delta z(\partial \rho/\partial t)$. Equating these two quantities leads to

$$\frac{\rho(z + \Delta z) u(z + \Delta z) - \rho(z) u(z)}{\Delta z} = -\frac{\partial \rho}{\partial t} \qquad (2.7)$$

Taking the limit as $\Delta z \to 0$ gives the differential form

$$\frac{\partial(\rho u)}{\partial z} + \frac{\partial \rho}{\partial t} = 0 \qquad (2.8)$$

Using the same reasoning here as presented above in Equation (2.6) regarding neglecting second-order terms, it can be shown (see Problem 2.1) that Equation (2.8) reduces to

$$\rho_0 \frac{\partial u}{\partial z} + \frac{\partial \rho_1}{\partial t} = 0 \qquad (2.9)$$

At this point it is necessary to consider the properties of the medium by introducing a relationship which tells how effectively the pressure field compresses the material to change its density (without loss or gain of heat).

2.1 DERIVATION OF THE WAVE EQUATION

This relationship defines the adiabatic compressibility constant K for the material, and it is given by

$$\frac{\rho_1}{\rho_0} = Kp \tag{2.10}$$

which shows that the normalized density change ρ_1/ρ_0 is proportional to the pressure p causing that change, with the proportionality constant being the compressibility K. The ratio ρ_1/ρ_0 is sometimes called the condensation s.

Note that in expressing the normalized density change in Equation (2.10) as being proportional to pressure with the coefficient K, we have assumed a linear relationship between density change and pressure and have ignored any nonlinear terms. These nonlinear terms will lead to interesting wave effects such as frequency doubling, but the effects are generally very small in tissues and have not yet been exploited in a clinical bioinstrument. Therefore, we shall not include nonlinear terms in the wave equation we are developing here.

Using Equation (2.10) in Equation (2.9) gives

$$\frac{\partial p}{\partial t} + \frac{1}{K} \cdot \frac{\partial u}{\partial z} = 0 \tag{2.11}$$

Equations (2.6) and (2.11) are two equations in two variables and represent the coupling that occurs between the pressure field and velocity field in the material:

$$\frac{\partial p}{\partial z} + \rho_0 \frac{\partial u}{\partial t} = 0 \tag{2.6}$$

$$\frac{\partial p}{\partial t} + \frac{1}{K} \cdot \frac{\partial u}{\partial z} = 0 \tag{2.11}$$

The constants in the equations are merely constants of the medium. Since both time and space derivatives are involved, it is reasonable to expect that the solutions will be traveling waves, as seen in the next section. The easiest way to solve for these waves is to combine Equations (2.6) and (2.11) into one equation, known as the wave equation, by taking the partial derivative of Equation (2.6) with respect to z:

$$\frac{\partial^2 p}{\partial z^2} + \rho_0 \frac{\partial^2 u}{\partial z \, \partial t} = 0 \tag{2.12}$$

and by taking the partial derivative of Equation (2.11) with respect to t:

$$\frac{\partial^2 p}{\partial t^2} + \frac{1}{K} \cdot \frac{\partial^2 u}{\partial t \, \partial z} = 0 \tag{2.13}$$

Since the order of the partial derivatives in the second term of either equation may be interchanged, Equations (2.12) and (2.13) may be combined to give

$$\frac{\partial^2 p}{\partial z^2} - \rho_0 K \frac{\partial^2 p}{\partial t^2} = 0 \tag{2.14}$$

which is the one-dimensional wave equation for the pressure p. A similar wave equation may be obtained for the velocity u (see Problem 2.2).

2.2 SOLUTIONS TO THE WAVE EQUATION

Rather than integrating the wave equation to get a general solution, it is more efficient to try some trial solutions suggested by our previous experience with waves. Two trivial solutions which are obvious are

$$p = 0$$

and

$$p = \text{constant}$$

A more interesting solution to try is

$$p = p_+ \cos(\omega t - kz) \tag{2.15}$$

where p_+ is an amplitude constant, ω is the angular frequency of the wave in radians per second, and k is the so-called propagation constant; ω and k will be discussed in detail later. Substitution of Equation (2.15) into the wave equation shows that it is indeed a valid solution provided that the following relationship always holds:

$$k^2 = \rho_0 K \omega^2 \tag{2.16}$$

In other words, the material will support a pressure wave of the form of Equation (2.15) for which the frequency ω and propagation constant k are related by Equation (2.16). Equation (2.16) is known as the dispersion equation, or Brillouin equation, and since the constants ρ_0 and K appear, it is specific to the medium supporting the wave.

Of course, Equation (2.15) is not the only solution to the wave equation. Another possibility is

$$p = p_- \cos(\omega t + kz) \tag{2.17}$$

which upon substitution also satisfies the wave equation, but again only if Equation (2.16) holds. A more general solution, with arbitrary phase, is

2.3 THE WAVE NATURE OF THE SOLUTIONS

$$p = p_+ \cos(\omega t - kz + \phi) \tag{2.18}$$

where ϕ is an arbitrary phase angle. Again the dispersion equation must apply for this wave to be a valid solution. A commonly seen solution,

$$p = p_+ \sin(\omega t - kz) \tag{2.19}$$

is merely a special case of Equation (2.18) with $\phi = -\pi/2$.

It appears from the above solutions that there might be a large number of wave shapes which will satisfy the wave equation. This is true. In fact, a general wave whose formula is *any* function f which contains time and space in the form of terms $(\omega t \pm kz)$ will be a valid solution. That is, a pressure wave of the general form

$$p = p_\mp f(\omega t \pm kz) \tag{2.20}$$

may be shown to satisfy the wave equation, constrained only by the dispersion relationship (see Problem 2.4). Thus, shapes such as pulses or square waves will propagate intact as acoustic plane waves in the lossless medium.

Table 2.1 summarizes some of the solutions to the one-dimensional wave equation and reiterates the dispersion equation which relates angular frequency ω and propagation constant k for acoustic waves.

2.3 THE WAVE NATURE OF THE SOLUTIONS

For the remainder of the discussion on waves we shall use Equation (2.15) as a typical example of a valid solution to the pressure wave equation. This choice of a sinusoidal wave form is not as restrictive as it may seem, for two reasons: (1) many sources emit ultrasound waves that are approximately sinusoidal in shape, such as narrowband or continuous-wave transducers, and (2) even an arbitrary wave shape, like a narrow time pulse, can be decomposed into its sinusoidal Fourier components, each of which will have a form like Equation (2.18) (with unique amplitude and phase). This possibility of breaking down a complex wave shape into its more basic sinusoidal components is assured because superposition is allowed in the solution of a linear equation, and Problem 2.7 shows that the one-dimensional wave equation is linear.

Looking therefore at Equation (2.15) we now determine the traveling characteristics (i.e., the time and space relationships) for such a wave. Figure 2.3a plots the magnitude of the pressure

$$p = p_+ \cos(\omega t - kz)$$

as a function of longitudinal distance z at one instant of time, $t = 0$. The oscillatory behavior with distance is obvious, with the spacing of the repeat

TABLE 2.1 SOME SOLUTIONS TO THE ONE-DIMENSIONAL WAVE EQUATION FOR THE PRESSURE VARIABLE p

Wave Equation

$$\frac{\partial^2 p}{\partial z^2} - \rho_0 K \frac{\partial^2 p}{\partial t^2} = 0$$

Some solutions

$p = 0$	Trivial
$p = $ constant	Trivial
$p = p_+ \cos(\omega t - kz)$	Wave traveling in $+z$ direction
$p = p_- \cos(\omega t + kz)$	Wave traveling in $-z$ direction
$p = p_+ \sin(\omega t - kz)$	Phase shifted by $\pi/2$
$p = p_+ \cos(\omega t - kz + \phi)$	Arbitrary phase
$p = p'[\cos(\omega t - kz) + \cos(\omega t + kz)]$	Combination of two traveling waves producing a standing wave
$p = p_\pm f(\omega t \pm kz)$	Any function f of $\omega t \pm kz$

Dispersion Equation
$$k^2 = \rho_0 K \omega^2$$

of any feature (say, the positive peaks, whose positions are indicated by dashed lines in the figure) being given by the distance required to make the argument or phase of the cosine term go through a change of 2π. Since the phase is $\omega t - kz$, and t is fixed here, a change of 2π in the phase is achieved whenever kz increases by 2π. This requires an increase of $2\pi/k$ in the distance z, and this repetition spacing is known as the *wavelength* λ of the wave. Thus,

$$\boxed{\lambda = \frac{2\pi}{k}} \qquad (2.21)$$

The dashed lines in Figure 2.3a are all drawn at positions of constant phase (equal to integral multiples of 2π for this example) and are often called constant *phase fronts*. Actually, these phase fronts are surfaces which are x-y planes perpendicular to the z axis. Because the argument of the cosine term does not vary with either x or y, the phase of the wave is constant throughout any entire x-y plane and is only dependent upon z. Waves with this feature are known as *plane waves*, and it can be seen that the one-dimensional wave equation given by Equation (2.14) leads to plane wave solutions.

Now consider what happens to Figure 2.3a when time is allowed to advance by an amount Δt. Figure 2.3b plots the wave as a function of distance, but at a later time, $t = \Delta t$. Again the oscillatory nature is evident, but the positions of the features of the wave have advanced. For example, the dashed lines representing the peaks or phase fronts have moved an

2.3 THE WAVE NATURE OF THE SOLUTIONS

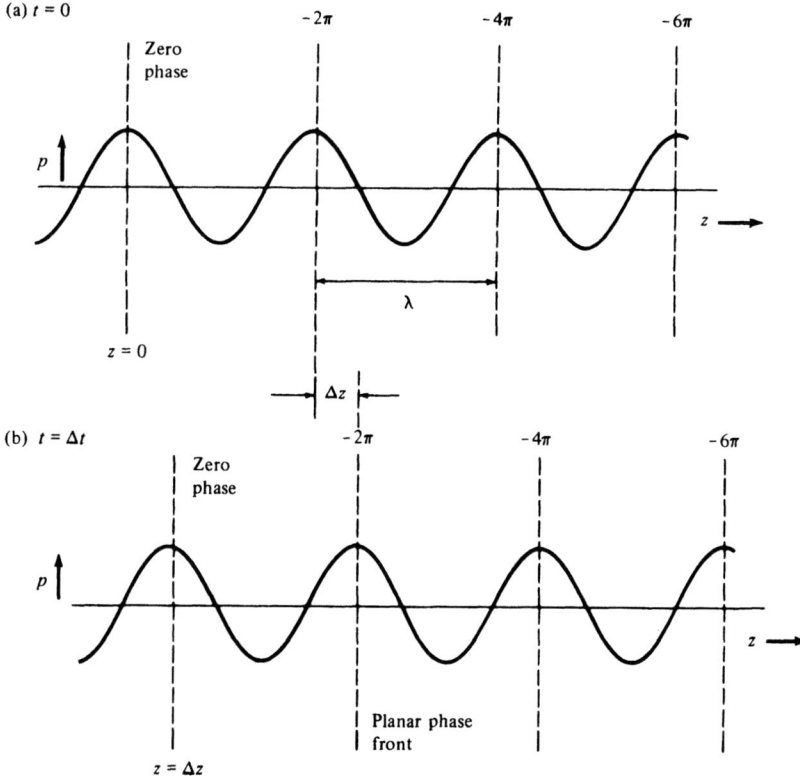

Figure 2.3 (a) The pressure wave as a function of distance at time $t = 0$; (b) at $t = \Delta t$. The planar phase fronts have advanced in amount Δz.

amount Δz. We can calculate the velocity of these phase fronts by solving for the distance Δz which the dashed line representing the phase front with zero phase has moved; for this front $(\omega t - kz)$ always equals zero. When $t = 0$, it is located at $z = 0$. When $t = \Delta t$, it is located at $z = \Delta z$ such that

$$\omega \, \Delta t - k \, \Delta z = 0$$

or

$$\frac{\Delta z}{\Delta t} = \frac{\omega}{k} \tag{2.22}$$

Equation (2.22) gives the velocity of the phase fronts and is known as the longitudinal *phase velocity* c of the wave:

$$\boxed{c = \frac{\omega}{k}} \tag{2.23}$$

The phase velocity is denoted by the letter c by convention in ultrasound (and optics) texts; its value will depend upon the medium supporting the waves.

A more general, but equivalent, derivation of the phase velocity for all our solutions to the wave equation may be obtained by examining the equation that specifies the condition for constant phase:

$$\omega t \pm kz + \phi = \text{constant} \tag{2.24}$$

In differential form, this equation is

$$\omega \, dt \pm k \, dz = 0$$

or

$$c = \frac{dz}{dt} = \frac{\mp \omega}{k} \tag{2.25}$$

By keeping track of the corresponding signs in Equations (2.24) and (2.25), it can be seen that whenever a negative sign occurs between terms ωt and kz in the phase argument of the wave solution, this formula represents a wave traveling in the positive z direction, as in Equation (2.15). Mathematically, as t increases, z must also increase to keep the phase constant. By similar reasoning, a positive sign in the argument, such as Equation (2.17), gives a wave traveling in the negative z direction (see Problem 2.5).

The *ray* associated with a plane wave is simply a vector pointing in the direction of propagation of the wave. For the plane wave considered in Figure 2.3, the ray is perpendicular to the phase fronts and is pointing in the $+z$ direction.

Not only does the solution plotted in Figure 2.3 display sinusoidal behavior as a function of distance z but also as a function of time t. If an observer were to be stationary in space at some position $z = z_0$, the wave would travel past with a sinusoidal time nature. The *temporal frequency* of the sinusoidal oscillation would be given by

$$f = \frac{\omega}{2\pi} \tag{2.26}$$

and by combining Equations (2.21), (2.23), and (2.26), the following very useful relationship can be obtained:

$$\boxed{\lambda = \frac{c}{f}} \tag{2.27}$$

Thus, if the frequency and phase velocity of a wave are known, the corresponding wavelength can easily be calculated. As a useful aid to remembering Equation (2.27), the simple picture shown in Figure 2.4 can be

2.3 THE WAVE NATURE OF THE SOLUTIONS

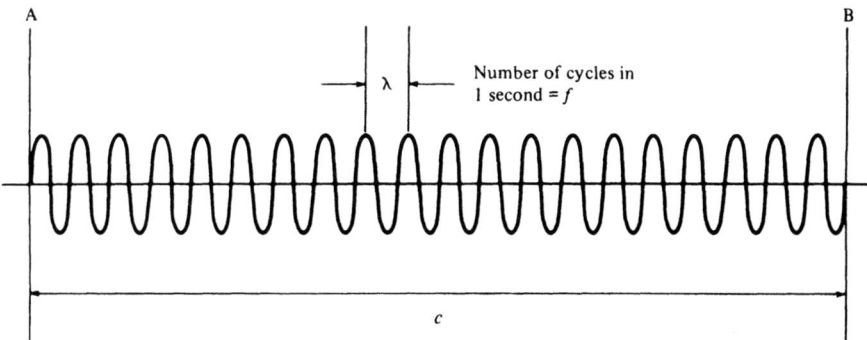

Figure 2.4 A simple picture showing a 1-second segment of a wave. From this, it can be seen that $\lambda = c/f$.

employed. It shows a length of wave that has propagated from an initial position A to position B during a 1-second time period. The length of travel in 1 second is obviously $c(1) = c$. The number of cycles contained in the 1-second sample is just the frequency f. Therefore, the length of one cycle (i.e., the wavelength λ) is found by dividing the total length by the number of cycles:

$$\lambda = \frac{c}{f}$$

which is identical to Equation (2.27).

It is interesting to point out here that all of the discussion and formulas derived so far in this section apply equally well to all types of waves: sound waves and electromagnetic waves such as light, microwaves, radio waves, etc. Where, then, does specialization to acoustic waves come into effect? The exact nature of the wave enters when the relationship between ω and k (the dispersion equation) is specified. For plane compressional acoustic waves, Equation (2.16) gives

$$\omega^2 = \left(\frac{1}{\rho_0 K}\right) k^2$$

From Equation (2.23) the phase velocity for these waves is

$$\boxed{c = \frac{\omega}{k} = \frac{1}{\sqrt{\rho_0 K}}} \qquad (2.28)$$

which is a function of the average density ρ_0 and compressibility K of the material in which the waves are propagating. The values for these constants may be readily obtained in handbooks for most materials. For example, water at 35°C possesses the following constants:

$$K = 4.48 \times 10^{-10} \quad m^2/N$$
$$\rho_0 = 10^3 \quad kg/m^3$$

From Equation (2.28),

$$\boxed{c_{water} \approx 1.5 \times 10^3 \quad m/s = 1.5 \times 10^5 \quad cm/s}$$

The reader will find that this important value will be used often in the estimation of wavelength and pulse arrival times in soft tissues, and it is profitably memorized.

We can now calculate the size of a wavelength of ultrasound in water for a typical frequency $f = 3$ MHz by utilizing Equation (2.27):

$$\lambda = \frac{c}{f} = \frac{1.5 \times 10^5 \quad cm/s}{3 \times 10^6/s} = 0.05 \quad cm$$

It will be seen in a later chapter that the phase velocity of ultrasound in most soft biological tissue is within ±5% of the value for water. Thus, at a given frequency, the velocity and wavelength in tissue will be similar to

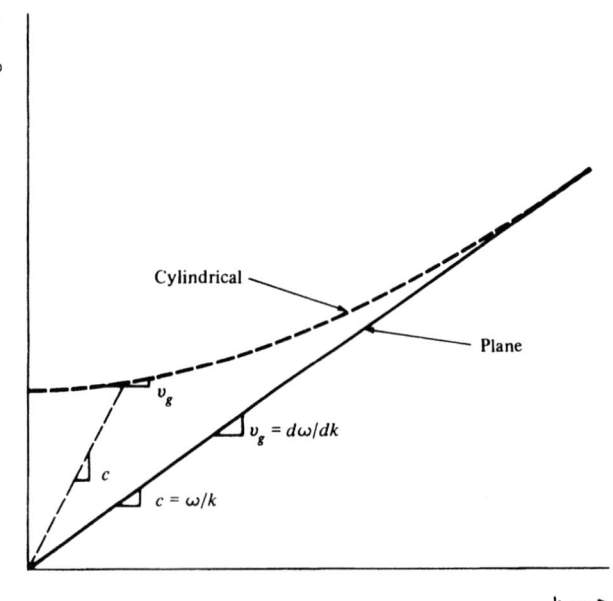

Figure 2.5 Plot of the dispersion equation for two types of waves. For plane waves (solid line), the ratio ω/k is a constant, and therefore the phase velocity is independent of frequency ("dispersionless" system). Certain other waves, however, such as those propagating inside cylindrical boundaries (dashed line), exhibit "dispersion" since their phase velocity ω/k is a function of frequency.

2.3 THE WAVE NATURE OF THE SOLUTIONS

TABLE 2.2 SOME USEFUL RELATIONSHIPS FOR WAVES

Waves in general

$$\lambda = \frac{2\pi}{k}$$

$$c = \frac{\omega}{k}$$

$$f = \frac{\omega}{2\pi}$$

$$\lambda = \frac{c}{f}$$

Plane acoustic waves

$$c = \frac{1}{\sqrt{\rho_0 K}}$$

For water: $c \simeq 1.5 \times 10^5$ cm/s

---

that in water. This small wavelength is the basic reason why ultrasound has good resolution in imaging instruments. An electromagnetic wave at the same frequency would have a much longer wavelength, due to the much higher phase velocity of electromagnetic waves in tissue (see Problem 2.9).

It is evident now that the dispersion equation plays a key role in specifying the wave characteristics for whatever system is under consideration. From Equations (2.16) and (2.28), the ratio ω/k is a constant for plane-compressional acoustic waves, and therefore the phase velocity is independent of frequency. This has been shown to be essentially true experimentally. Such a situation is known as a *dispersionless* system. However, for other types of waves such as nonplanar waves confined within boundaries (e.g., waves inside a cylindrical waveguide), the phase velocity is not constant with frequency; such a system displays *dispersion*.

Figure 2.5 shows how such behavior can be graphically displayed by plotting the dispersion equation with ω as a function of k. On this diagram—known as a dispersion diagram, or Brillouin diagram—the phase velocity ω/k is given by the slope of a line that extends from the origin and intercepts the equation line at the frequency of operation. For plane waves, the dispersion equation plots as a straight line through the origin, so the phase velocity ω/k is constant with frequency. In our study, we will consider only plane waves or near-planar waves with no dispersion, since this is a good approximation to most physical situations in imaging.*

---

* Actually, when the propagation of an ultrasound *pulse* is of interest, it is more correct to consider the propagation of the pulse envelope, given by the group velocity $v_g = d\omega/dk$, which is the local slope of the dispersion equation, rather than by the phase velocity. However, for the plane waves considered in this text, the system is dispersionless and $c = \omega/k$ is a constant with respect to frequency, leading to the result $c = v_g$ (see Figure 2.5).

Table 2.2 summarizes some of the important wave relationships just derived.

PROBLEMS

2.1. Using the approximations described in the text, derive Equation (2.6) from Equation (2.5), and derive Equation (2.11) from Equation (2.8).

2.2. From Equations (2.6) and (2.11), derive the wave equation for the velocity variable u.

2.3. Find one nontrivial solution for the velocity u that satisfies the *velocity* wave equation derived in Problem 2.2, and by substitution find the dispersion equation that must hold for this solution to be valid.

2.4. (a) Show that the general formula for a pressure wave given by

$$p = p_{\mp} f(\omega t \pm kz)$$

where f is any function in which t and z appear only in terms of the form $\omega t \pm kz$, is a solution to the wave equation.

(b) Choose one example of a nonsinusoidal wave shape and give its mathematical formula $f(\omega t \pm kz)$. Draw its shape as a function of distance at two instants of time, t and $t + \Delta t$, and derive from your drawing the phase velocity for this wave in terms of ω and k. [*Hint:* the Dirac delta function $\delta(\omega t \pm kz)$ is an especially easy function to use for this purpose.]

2.5. Show in a manner similar to Figure 2.3 that $p = p_- \cos(\omega t + kz)$ represents a plane wave propagating in the negative z direction.

2.6. Consider a pressure wave whose formula is

$$p = 10 \cos(10^7 t + 10^2 z) \quad \text{N/m}^2$$

with t given in seconds and z in centimeters. Find ω, k, f, λ, c, and direction of propagation. Be sure to include correct units for each quantity. (*Note:* In SI units, N/m² is equivalent to the unit Pa, called pascal.)

2.7. Prove that the wave equation given by Equation (2.14) is a linear equation with regard to pressure p by showing that superposition holds for this equation (i.e., if f_1 and f_2 are both separate solutions, then $f_1 + f_2$ is also a solution).

2.8. Based upon your everyday experiences, rank the materials air, water, iron, and rubber in order of:
  (a) Increasing compressibility K.
  (b) Increasing density ρ_0.

2.9. (a) Calculate the wavelength of a 1-MHz ultrasonic wave in tissue, assuming that its phase velocity in tissue is close to that of water.
  (b) Calculate the wavelength of a 1-MHz electromagnetic wave in tissue, assuming that its phase velocity in tissue is about one-tenth of that in free space.

2.10. The compressibility of methyl alcohol is approximately 13×10^{-10} m²/N, and its density is approximately 0.78 g/cm³. Write the dispersion equation for an ultrasonic plane wave propagating in methyl alcohol, and plot this equation (ω vs. k) for a range of frequencies from 0 Hz to 10 MHz, similar to Figure 2.5. Then, determine from this figure the numerical values of:

(a) The phase velocity at a frequency of 1 MHz.
(b) The phase velocity at a frequency of 10 MHz.
(c) The propagation constant k at a frequency of 10 MHz.
(d) The wavelength at a frequency of 10 MHz.

Chapter 3

Impedance, Power, and Reflection

3.1 INTRODUCTION

In Chapter 2 the wave equation for the acoustic pressure p was derived by combining a pair of coupled equations, Equations (2.6) and (2.11), each relating p and particle velocity u, into a single wave equation containing only the dependent variable p. Just as easily an identical wave equation containing only the variable u could have been developed from that original pair of equations (see Problem 2.2), and solutions for the velocity u could have been obtained. Since the same pair of equations leads to solutions for both p and u, it is natural to expect that some relationship always exists in the medium between the quantities p and u. This is true, as shown next.

3.2 IMPEDANCE OF A MEDIUM

We let the relationship between the values of p and u be defined by the specific acoustic impedance Z as follows:

$$Z = \frac{p}{u} \qquad (3.1)$$

It is assumed that this relationship holds for all points in the medium at all instances of time. In general, Z may be a complex number whose phase angle represents the difference in phase between p and u; however, for the lossless plane waves considered so far, Z will be shown to be real.

We can find an expression for Z in terms of the properties of the medium by recalling one of the pair of original coupled equations relating p and u, Equation (2.6):

$$\frac{\partial p}{\partial z} + \rho_0 \frac{\partial u}{\partial t} = 0 \qquad (2.6)$$

Putting Equation (3.1) into Equation (2.6) gives

$$\frac{\partial p}{\partial z} + \frac{\rho_0}{Z} \frac{\partial p}{\partial t} = 0 \qquad (3.2)$$

since Z is a constant. One solution for p used in the last chapter,

$$p = p_+ \cos(\omega t - kz)$$

can be substituted into Equation (3.2) to give

$$kp_+ \sin(\omega t - kz) - \frac{\omega \rho_0}{Z} p_+ \sin(\omega t - kz) = 0$$

or, by canceling common terms,

$$k - \frac{\omega \rho_0}{Z} = 0$$

$$\boxed{Z = \frac{\omega \rho_0}{k} = \rho_0 c} \qquad (3.3)$$

This is the basic equation from which the impedance of a material may be found. Since the value of $\rho_0 c$ is characteristic of a particular material, the quantity $\rho_0 c$ is sometimes called its *characteristic* impedance. A special type of rubber whose impedance matches that of water to avoid reflections when the rubber is used to line ultrasound water tanks and apparatus has been named "rho-c" rubber.

Another form for impedance can be obtained by substituting Equation (2.28), which relates phase velocity to material properties, into Equation (3.3):

$$Z = \rho_0 \left(\frac{1}{\sqrt{\rho_0 K}} \right) = \sqrt{\frac{\rho_0}{K}} \qquad (3.4)$$

Equations (3.3) and (3.4) both show that the impedance of a medium depends upon its mechanical properties, and it will vary from material to material. Since we have considered only the lossless plane wave equation,

3.2 IMPEDANCE OF A MEDIUM

the impedance is also a real number. For water at 35°C, we can use the handbook numbers for ρ_0 and K given in the last chapter to get

$$Z_{water} = \sqrt{\frac{10^3 \text{ kg/m}^3}{4.48 \times 10^{-10} \text{ m}^2/\text{N}}} = 1.5 \times 10^6 \text{ kg/m}^2 \text{ s}*$$

The values of the impedance for biological tissues can be obtained in a similar fashion; in a later chapter they will be shown to be close to that for water.

Even though Equations (3.3) and (3.4) were derived by using a particular sinusoidal solution for p, any of the valid solutions for p found in the last chapter will lead to the same impedance equations. Thus, the above expressions for impedance are general. In addition, although Equation (2.6) was chosen as the starting equation for the derivation above, the other member of the original pair of coupled equations, Equation (2.11), could alternatively be used at the start, leading to an identical result for impedance (see Problem 3.1).

With impedance now defined, expressions for the particle velocity u can be obtained immediately, corresponding to any solution for the pressure p. For example, if the pressure wave is

$$p = p_+ \cos(\omega t - kz)$$

then the particle velocity wave accompanying this pressure is

$$u = \frac{p_+}{Z} \cos(\omega t - kz) = u_+ \cos(\omega t - kz) \tag{3.5}$$

Note that the pressure wave and the velocity wave are always in phase for the case of lossless plane waves.

Also note the following important fact: Since Equation (3.5) for the particle velocity has a sinusoidal form, the particles in the medium oscillate with equal positive and negative excursions about zero. That is, there is no *net* movement of the material's particles participating in the wave motion; the particles (atoms or molecules) merely vibrate back and forth with frequency f around their equilibrium position. Even though the wave phase fronts are progressing in some direction with phase velocity c, the particles themselves are oscillating in place. From this it can be seen that the phase velocity c and particle velocity u are two distinctly different quantities and should not be confused.

How is it that a wave can propagate over perhaps large distances with a phase velocity c and yet the particles in the medium exhibit no net motion? To resolve this paradox, consider the analogy of students standing in a

---

* Equivalent impedance units are $\text{kg/m}^2\text{s} \equiv \text{sN/m}^3 \equiv \text{sPa/m} \equiv$ "rayl".

long line waiting to buy tickets for a rock concert. At the very end of the line, one student shoves his neighbor ahead of him. In turn, this student shoves his neighbor toward the front, and so on, the shove progressing from student to student all the way to the front of the line. Although each student does not vary in average position, the effect of the shove is propagated down the entire length of the line. The velocity of propagation of the shove will be dependent in some way upon the mass of each student and the speed of reaction of each participant. The same picture would apply to any vibratory shaking that might be initiated by the student at the end of the line. (Whether or not the shoving wave would be reflected at the *front* of the line depends upon the interface between the student first in line and the ticket booth. If the ticket booth is either very rigid or collapses entirely, the front-end student may understandably turn around and start a shove going backward down the line. However, if the ticket booth yields gently in accordance with the student, he might forget it and not reflect the wave. A later part of this chapter contains further discussion about reflections.)

3.3 POWER DENSITY

Now that expressions have been found for both particle velocity and pressure, we can derive an expression for the power density carried by the wave. The instantaneous power flowing through a unit area perpendicular to the direction of propagation of the wave as one elemental volume of the fluid acts on a neighboring element is defined as the wave's *power density* and is given by

$$\text{Power density} = \frac{\text{power}}{\text{area}} = \frac{\text{work}}{\text{area} \cdot \text{time}} = \frac{\text{force} \cdot \text{distance}}{\text{area} \cdot \text{time}} = \text{pressure} \cdot \text{velocity}$$

or

$$I = pu \tag{3.6}$$

So, using Equation (3.1),

$$\boxed{I = \frac{p^2}{Z} = Zu^2} \tag{3.7}$$

The total power of a uniform wave is given by the product of its power density I and its cross-sectional area A.

In Figure 3.1, I is plotted as a function of distance for a sinusoidal pressure wave at one instant of time. Note that I is always a positive quantity and the spacing between power peaks is one-half a pressure wavelength.

3.3 POWER DENSITY

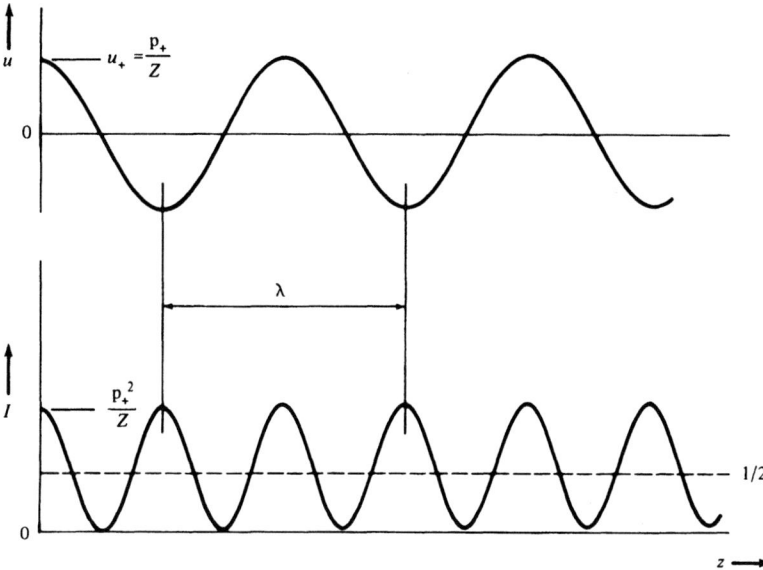

Figure 3.1 Variation of particle velocity u and power density I corresponding to a pressure wave of the form $p = p_+ \cos(\omega t - kz)$. I has an average value equal to one-half its peak value.

As a practical matter it is often more convenient to consider the space average of the power density. The space average of I for a sinusoidal acoustic wave $p = p_+ \cos(\omega t - kz)$ can be found by simply integrating the instantaneous power density over one wavelength, then dividing by one wavelength:

$$I_{\text{ave}} = \frac{1}{\lambda} \int_0^\lambda \frac{p^2}{Z} dz = \frac{k}{2\pi} \int_0^{2\pi/k} \frac{p_+^2}{Z} \cos^2(\omega t - kz) dz = \frac{kp_+^2}{2\pi Z} \cdot \frac{\pi}{k} = \frac{p_+^2}{2Z} \quad (3.8)$$

Thus, the average power density (or *intensity*) for a sinusoidal wave is just one-half of its peak value, as might be expected from an examination of Figure 3.1. A time average of I will yield an identical result (see Problem 3.4).

Equations (3.7) and (3.8) show that doubling the magnitude of pressure will quadruple the power in the wave. This is analogous to the relationship between voltage and power in an electrical circuit.

3.4 REFLECTION OF WAVES AT INTERFACES

Imaging instrumentation based upon the penetration of ultrasound into complex objects would generally not be possible were it not for the reflection of the waves revealing the various internal interfaces of the object. As will be shown in this section, whenever a wave passes from a region of one value of acoustic impedance into a neighboring region of different impedance, a certain amount of the incident power is reflected at the boundary and the remainder continues as a transmitted wave. The reflected wave serves as an indicator of the boundary position and shape, whereas the transmitted portion probes deeper interfaces. A study of the nature of ultrasonic reflection is therefore basic to a technical understanding of imaging instrumentation.

There are two important items to be considered when studying the reflection of waves. The first deals with the angle that the reflected wave has as it leaves the interface and the angle that the transmitted wave takes as it propagates into the new region. The second is the percentage of incident power that is reflected at the boundary. These items will be discussed in order.

3.4.1 Angles of Reflection and Transmission

The situation that occurs when an incident plane wave impinges upon a plane boundary between two regions with different phase velocities is diagrammed in Figure 3.2. In response to the vibrations set up by the incident wave at the surface of the boundary, two additional waves are produced: a reflected wave and a transmitted (refracted) wave. These additional waves are constrained both in angle and magnitude since they must match the excitations of the incident wave (their source) along the boundary. For example, consider the incident wave whose phase fronts are shown on the left of the interface in Figure 3.2 by solid lines as it strikes the boundary at an angle θ_i between its direction of propagation (ray direction) and a normal to the plane of the boundary. The perpendicular spacing between the phase fronts is defined in medium 1 as the wavelength λ_1, but note that the spacing between these phase fronts along the *boundary* (designated

3.4 REFLECTION OF WAVES AT INTERFACES

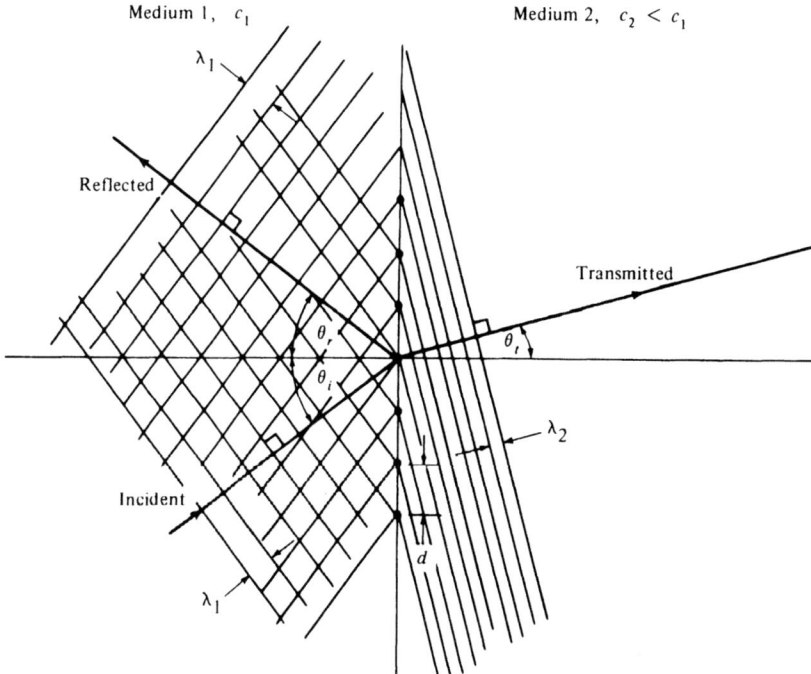

Figure 3.2 All three waves have angles such that the projections of their phase fronts onto the boundary (given by the distance d) are equally spaced. This leads to the law of specular reflection and Snell's law. For the case shown here, $c_2 < c_1$ so that $\theta_t < \theta_i$.

by the distance d) is not equal to λ_1. When the wave strikes the interface at an angle, the distance between phase fronts along the interface will be larger than λ_1, given by simple geometry as

$$d = \frac{\lambda_1}{\sin \theta_i} \quad (3.9)$$

Now, since the vibrations set up by the incident wave along the boundary cause the radiation of both the reflected and the transmitted waves, the projected distance between the phase fronts of these waves must exactly match those of the incident wave along the entire interface. The reflected wave, whose phase fronts are indicated in Figure 3.2 by solid lines, has the same wavelength in medium 1 as the incident wave because the frequency is the same for all three waves, and $\lambda = c/f$ from Chapter 2. It is therefore easy to see that in order to match the spacing d, the reflected wave must be radiated back at a reflection angle equal to the angle of incidence, or

$$\theta_r = \theta_i \quad (3.10)$$

This is the law of specular, or "mirrorlike," reflection and is consistent with our everyday experience with light waves, such as the optical reflections from a smooth pond of water.

The transmitted wave, shown in Figure 3.2 by solid lines on the right of the interface, has a different wavelength in medium 2 than the waves in medium 1. This is true because $\lambda = c/f$ and $c_1 \neq c_2$. Thus,

$$\lambda_1 \neq \lambda_2 \tag{3.11}$$

In order for the transmitted wave to match the spacing of its phase fronts along the boundary to d, it must be tilted at an angle different from the angle of the incident wave. Again, geometry shows (see Problem 3.6) that the spacing d will be matched when

$$\boxed{\frac{\sin \theta_i}{\sin \theta_t} = \frac{\lambda_1}{\lambda_2} = \frac{c_1}{c_2}} \tag{3.12}$$

Equation (3.12), known as Snell's law, applies as well to plane electromagnetic waves as it does to plane acoustic waves. It shows that whenever a wave strikes a boundary between two media with different phase velocities at a nonperpendicular angle, the transmitted wave will propagate at an angle that is different from the angle of incidence. This phenomenon is known as *refraction* of the transmitted wave and the ratio on the right side of Equation (3.12) is known as the index of refraction of material 2 with respect to material 1. Note for the special case when $\theta_i = 0$ (normal incidence), $\theta_t = 0$, and there is no refraction in this case regardless of the ratio of phase velocities in the two regions.

Snell's law, as presented in Equation (3.12), gives the precise angle of refraction for any combination of materials forming the interface. In many situations, however, one needs only to quickly discern the direction of refraction rather than the exact angle, and some means of rapidly remembering the direction of bending of the wave propagation would be convenient. Figure 3.3 presents a simple memory aid for such a purpose. Think of the wave's phase fronts as parallel columns of marching soldiers, each marching shoulder to shoulder in a parade. All soldiers are marching in rhythm, but as each line passes from medium 1 into medium 2, the speed of walking changes as the stride increases or decreases proportionally to the relative phase velocity of the region. If the lines are entering the boundary at an angle, one side of each line changes its speed sooner than the opposite side, bending the lines. For example, consider a parade entering a slower region from a faster region, as in Figure 3.3a. The lines will bend such that the parade propagates with an angle of transmission smaller than the angle of incidence. Figure 3.3b shows the opposite case. When the parade passes from a region of slower velocity into a region of faster velocity, the parade will bend further away from the normal direction. These simple

3.4 REFLECTION OF WAVES AT INTERFACES

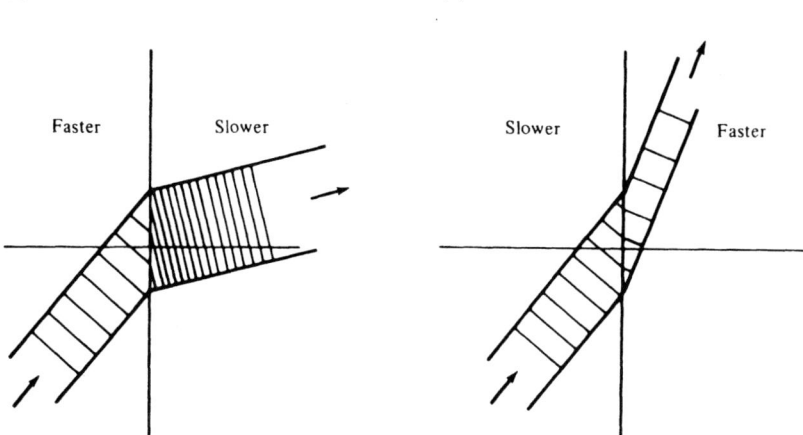

Figure 3.3 A memory aid giving the direction of change in the angle of propagation when a wave passes (a) from a faster to slower region and (b) from a slower to a faster region.

pictures predict the correct refraction direction as given by Snell's law. Waves of ultrasound behave in the same manner.

3.4.2 Magnitudes of Reflected and Transmitted Waves

We have shown that the angles of reflection and transmission of a wave may be calculated by Snell's law. But nothing has been mentioned yet about the percentage of power of the incident wave that is converted into the power of the reflected and the transmitted waves. A new consideration, namely, the continuity of pressure and normal velocity components across the interface, must be invoked to give these quantities. Figure 3.4 shows the pressures and particle velocity vectors which are associated with each wave immediately adjacent to the interface between medium 1 and medium 2. Since the pressures of each wave act in all directions, and since the boundary itself does not supply any resistance, the net pressure on the left side of the boundary must equal the net pressure on the right side of the boundary. This fact gives a relationship between the three pressures, $p_t = p_i + p_r$, but does not yet uniquely fix them. It is also required that the net velocity components of the particles perpendicular to the left boundary side must be equal to the perpendicular velocity component on the right side. Otherwise the media would pull apart from each other at the boundary. Thus, $u_t \cos \theta_t = u_i \cos \theta_i - u_r \cos \theta_r$. This requirement on the velocities may be related to the respective pressures through the impedance equation,

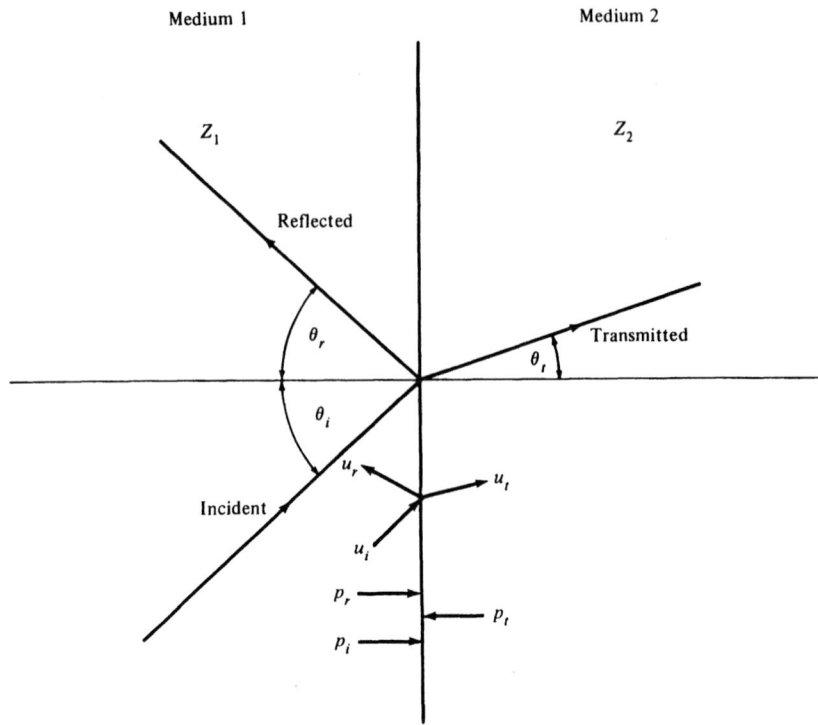

Figure 3.4 The total pressure and the normal components of particle velocity must be continuous across the boundary. This leads to Equation (3.13) for the reflection coefficient R.

Equation (3.1), thereby adding enough information to the model to uniquely specify the ratios of reflected pressure to incident pressure.

Note that the above boundary conditions have not attempted to match wave values tangential to the boundary. This is impossible to do using only the compressional-type (longitudinal) waves of our model—it requires the addition of shear waves. Reflected and transmitted shear waves will therefore generally be set up in addition to reflected and transmitted compressional waves by an incident compressional wave when it strikes a boundary at a nonzero angle. This phenomenon is known as *mode conversion*. However, except for propagation in bone, shear waves decay very rapidly, and only the compressional waves remain at distances useful for medical imaging.

We define the ratio of reflected pressure to incident pressure to be the *reflection coefficient R*. As might be expected from the discussion above, this coefficient is a function of the acoustic impedances of the two regions.

3.4 REFLECTION OF WAVES AT INTERFACES

By requiring continuity of the pressures and normal velocities, a mathematical expression for R is obtained (see Problem 3.10):

$$R = \frac{p_r}{p_i} = \frac{(Z_2/\cos\theta_t) - (Z_1/\cos\theta_i)}{(Z_2/\cos\theta_t) + (Z_1/\cos\theta_i)} \quad (3.13)$$

Equation (3.13) holds for any angle of incidence θ_i. The transmitted angle θ_t corresponding to θ_i may be obtained from Snell's law.

Since $p = Zu$, it is easy to show that the reflection coefficient may also be written as a ratio of particle velocities:*

$$R = \frac{p_r}{p_i} = \frac{Z_1 u_r}{Z_1 u_i} = \frac{u_r}{u_i} \quad (3.14)$$

Equation (3.13) reduces to a much simpler expression for the special case of a wave incident normal to the boundary. This case is often used to estimate the approximate power of reflection and transmission at interfaces internal to the body, although it is only an approximation since the actual internal tissue borders are rarely exactly perpendicular to the direction of propagation of the ultrasound pulse. However, the variation of R with respect to θ_i is slight in the region near $\theta_i = 0$ (see Problem 3.12) and the expression reduces to a simpler form for normal incidence. Let $\theta_i = \theta_t = 0$, and Equation (3.13) reduces to

$$R = \frac{Z_2 - Z_1}{Z_2 + Z_1} \quad (3.15)$$

Two interesting points can be seen by examining Equation (3.15). First, when a wave passes from one region into another with the same acoustic impedance, $Z_1 = Z_2$ and $R = 0$, so there is no reflected wave.† Thus, an impedance mismatch is needed at a boundary to produce any reflection, and the equation shows that the amount of reflected pressure is proportional to the normalized mismatch between the two regions. As the inequality between the two impedances increases, the amount of reflection will increase.

Second, the equation shows that when $Z_2 > Z_1$, R will be a positive number, whereas when $Z_2 < Z_1$, R will be negative. What does a negative

---

* The convention used in this text is that the direction of positive particle velocity for the reflected wave u_r is opposite to that for the incident wave u_i (refer to Figure 3.4). As a consequence, the ratio of particle velocities has the same sign as the ratio of pressures.

† This statement is strictly true only for the case of normal incidence. The more general Equation (3.13) used in conjunction with Snell's law shows that R will be nonzero for the case of $\theta_i \neq 0$ and $c_1 \neq c_2$, even though $Z_1 = Z_2$. But it is rare for the impedances of two materials to be equal if their phase velocities are unequal.

value for R represent? It merely means that for this situation, the *phase* of the reflected pressure wave will be 180° from the phase of the incident pressure wave at the interface. This is necessary in order to match the conditions of continuity at the boundary when $Z_2 < Z_1$. (As will be shown in the next section, the power of the reflected wave is proportional to R^2, which will always be a positive number as expected.)

Figure 3.5a shows an example of how an incident acoustic wave with pressure p_i and particle velocity u_i is divided into reflected and transmitted waves at the interface between two media with impedances Z_1 and Z_2, respectively. For the example shown, $Z_2 > Z_1$ so that R is a positive number and p_i and p_r are in phase at the interface; that is, at those times when p_i is positive, p_r is also positive. From the requirement of the continuity of pressure and the continuity of normal velocity components at the boundary, it can be seen that $p_t = p_r + p_i$ and $u_t = u_i - u_r$ for the case $\theta_i = 0$.

Figure 3.5a shows only one instant of time. As time advances, the waves will move in the directions shown by the arrows, but the boundary conditions given above will always hold at the interface. A similar figure could be drawn for the case of $Z_2 < Z_1$ (see Problem 3.13), in which circumstance R will be a negative number and p_i and p_r will have opposite signs at the interface. Still, the boundary conditions given above must always be satisfied by the three waves.

Standing Waves

An interesting effect occurs on the left side of the boundary due to the combination in space and time of the incident and reflected waves that are traveling in opposite directions. The resultant superposition of the two waves (shown in Figure 3.5b) forms a standing wave pattern described by an envelope possessing periodic maxima and minima within which the single resultant wave is always contained. The term "standing wave" is derived from the fact that the envelope is stationary in space; the positions of the envelope's maxima and minima do not propagate.

When the relative amplitudes of the incident and reflected waves are unequal, the standing wave envelope has only partial minima which do not reach zero, as shown in the example of Figure 3.5b. However, for the case of equal amplitudes for the incident and reflected waves such as occurs when the absolute value of the reflection coefficient is equal to unity, the standing wave envelope has complete minima where the resultant wave is always zero.

We can obtain an understanding of the nature of standing waves by examining the case when the reflected wave amplitude equals the incident wave amplitude. Mathematically, this situation can be described as the

3.4 REFLECTION OF WAVES AT INTERFACES

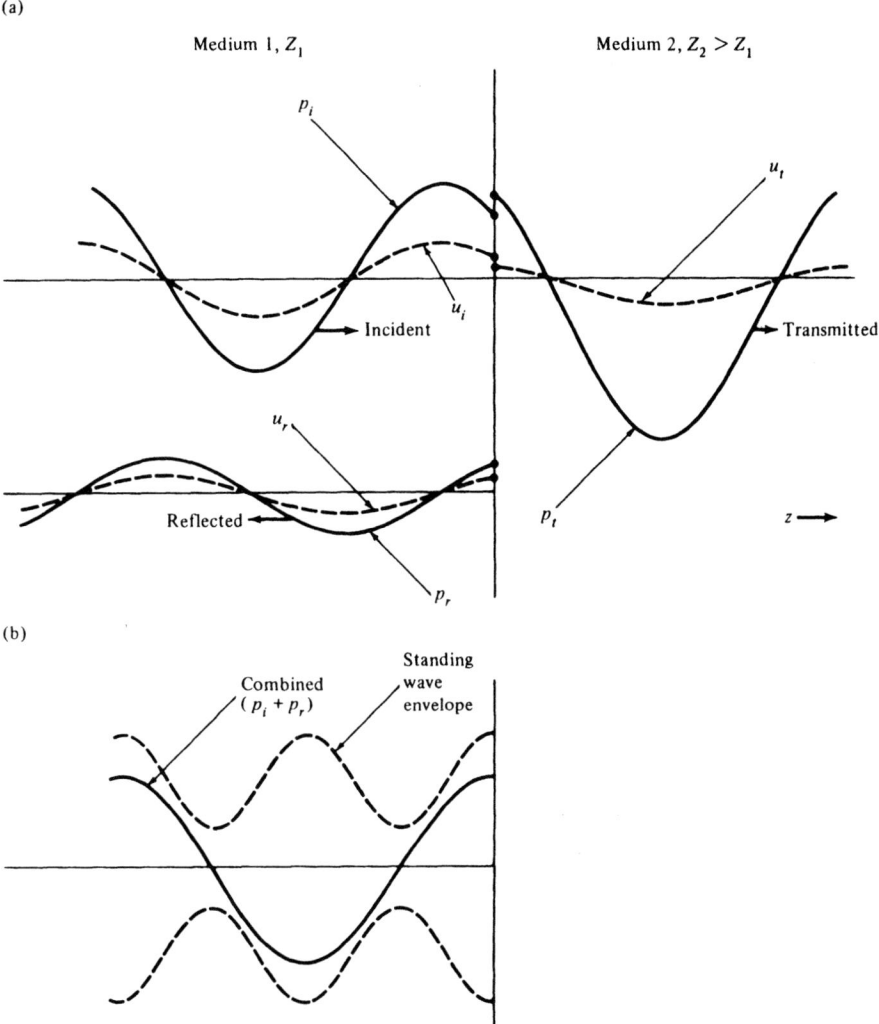

Figure 3.5 (a) An example of a wave with pressure p_i and particle velocity u_i striking an interface at normal incidence, resulting in transmitted and reflected waves. In the case shown, $Z_2 > Z_1$, so p_i and p_r are in phase at the interface. (b) The incident wave and reflected wave combine to form a partial "standing wave" in the region to the left of the boundary.

sum of two equal-amplitude sinusoidal waves, one traveling in the $+z$ direction and the other traveling in the $-z$ direction:

$$p = p_0[\cos(\omega t - kz) + \cos(\omega t + kz)] \qquad (3.16)$$

Using the trigonometric identity $\cos(A - B) + \cos(A + B) = 2 \cos A \times \cos B$ leads to

$$p = 2p_0 \cos(\omega t) \cos(kz) \qquad (3.17)$$

A time sequence of this resultant pressure is shown in Figure 3.6. Since $\cos(\omega t)$ alternates between $+1$ and -1 as a function of time, the envelope of the resultant wave has a spatial variation given by $\pm 2p_0 \cos(kz)$, shown at the bottom of Figure 3.6. The resultant wave is fixed in space yet oscillates in time. No net power is propagated.

It is rare in soft tissues alone to have impedance mismatches large enough to make R approach unity, but such mismatch is possible for tissue/air interfaces or tissue/bone interfaces. For example, consider an acoustic wave traveling in water striking a perpendicular interface with air. To calculate the reflection coefficient from Equation (3.15), the impedance of air must be calculated. From Equation (3.4)

$$Z_{air} = \sqrt{\frac{\rho_0}{K}}$$

Using handbook values for air at 35°C and 1 atm, $\rho_0 = 1.15$ kg/m^3 and $K = 9.9 \times 10^{-6}$ m^2/N, so

$$Z_{air} = \sqrt{\frac{1.15 \quad \text{kg/m}^3}{9.9 \times 10^{-6} \quad \text{m}^2/\text{N}}} = 341 \text{ kg/m}^2 \text{ s}$$

Since $Z_{water} = 1.5 \times 10^6$ kg/m^2 s from a previous section, Equation (3.15) gives

$$R = \frac{341 - 1.5 \times 10^6}{341 + 1.5 \times 10^6} \approx -1$$

Thus, it is seen that almost all incident acoustic power is reflected from a water/air interface and a strong standing wave pattern in the region of the water exists. Very little power is transmitted into the air cavity. This result leads to severe limitations of ultrasonic imaging of certain regions of the body such as the lungs.

3.4.3 Powers of Reflected and Transmitted Waves

The reflection coefficient R was defined as a ratio of pressures. We often are more interested in reflected power, so the relationship between pressure and power density, Equation (3.7), may be used to give the following ratio, valid for any angle of incidence:

$$\frac{\text{Reflected power density}}{\text{Incident power density}} = \frac{I_r}{I_i} = \frac{p_r^2/Z_1}{p_i^2/Z_1} = R^2 \qquad (3.18)$$

3.4 REFLECTION OF WAVES AT INTERFACES

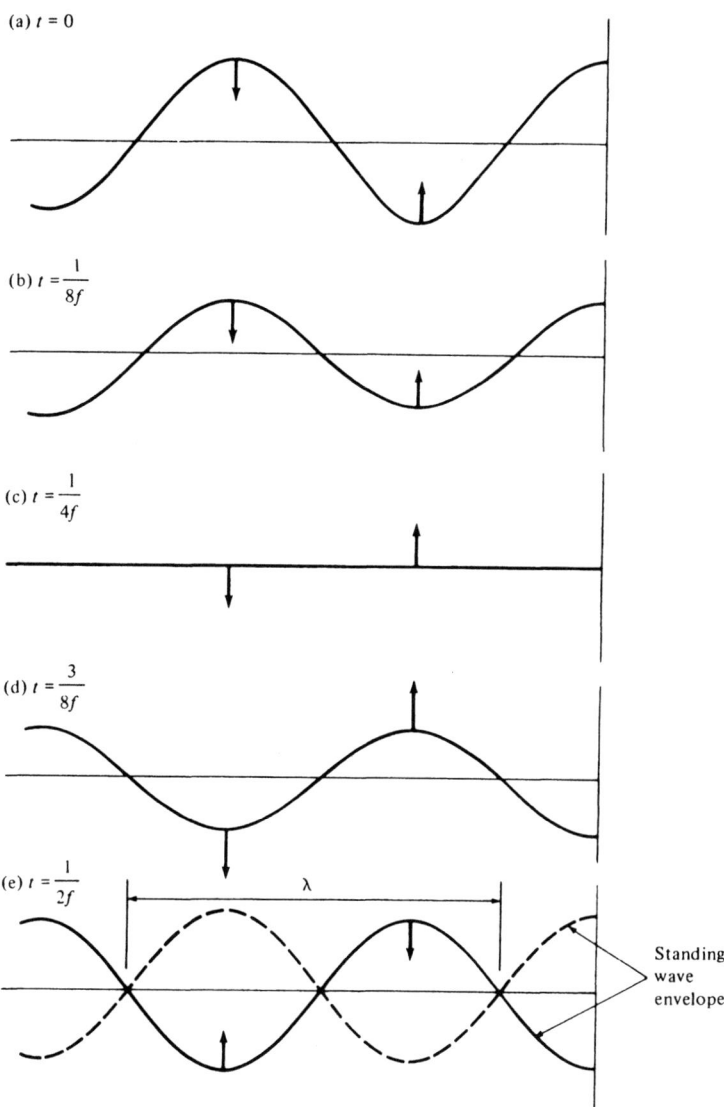

Figure 3.6 The time sequence of a standing wave produced by total reflection of the incident pressure wave from a boundary. The standing wave is the result of two equal-amplitude but countertraveling waves.

Thus, if Z_1 and Z_2 are known, the amount of reflected power can be found from Equations (3.13) and (3.18). Since the total power carried by a beam is just the product of the beam's power density I and its cross-sectional area A and since the areas of the incident beam and reflected beam are equal because $\theta_i = \theta_r$, the above equation also applies to the ratio of the total power of the two beams.

Also of interest is the amount of power density remaining in the transmitted portion of the wave after passage through the boundary. The power densities of the incident and transmitted waves are given by Equation (3.7):

$$I_i = \frac{p_i^2}{Z_1} \tag{3.19}$$

$$I_t = \frac{p_t^2}{Z_2} \tag{3.20}$$

Using the boundary condition $p_t = p_i + p_r$ and Equations (3.19) and (3.20) gives the following ratio:

$$\frac{\text{Transmitted power density}}{\text{Incident power density}} = \frac{I_t}{I_i} = \frac{[1+(p_r/p_i)]^2 Z_1}{Z_2} = (1+R)^2 \frac{Z_1}{Z_2} \tag{3.21}$$

Is this relationship consistent with the principle of the conservation of energy? That principle states that the total amount of power entering the boundary must equal the total amount of power leaving. Thus,

$$I_i A_i = I_t A_t + I_r A_r \tag{3.22}$$

where A_i, A_t, and A_r are the cross-sectional areas of the incident, transmitted, and reflected beams, respectively. Dividing Equation (3.22) by $I_i A_i$ and using Equation (3.18) and the fact that $A_i = A_r$, we get the ratio

$$\frac{\text{Transmitted beam power}}{\text{Incident beam power}} = \frac{I_t A_t}{I_i A_i} = 1 - \frac{I_r}{I_i} = 1 - R^2 \tag{3.23}$$

Problem 3.19 shows that Equation (3.23) derived from the conservation of total power is equivalent to Equation (3.21) for the ratio of power densities only when the incident beam is normal to the boundary, that is, when $\theta_i = \theta_t = 0$.* The reason for this is that although the conservation of power must always hold, power *density* is not conserved when the cross-sectional area of the beam changes as the beam crosses the boundary. For nonnormal angles of incidence, Figure 3.3 shows that the area of the beam will be altered as it passes across the interface, so that $A_t \neq A_i$.

In summary, if one is interested in calculating the power *density* in a transmitted beam for an arbitrary angle of incidence, Equation (3.21) should be used. If $\theta_i = 0$, Equation (3.23) may be used and will give the

---

* It is also possible for the two equations to be consistent if θ_i and θ_t are equal but nonzero. But this can only happen if $c_1 = c_2$, a rare occurrence under the prevailing assumption that $Z_1 \neq Z_2$.

3.4 REFLECTION OF WAVES AT INTERFACES

same answer as Equation (3.21). If one is interested in the *total* power carried by the transmitted beam, Equation (3.23) should be used for any angle of incidence.

Note that Equation (3.13) assures that $-1 \leq R \leq +1$, so the magnitude of R can never exceed unity. Thus, Equation (3.23) will always give a positive value for the transmitted power, as a confirmation of our intuition.

Decibel Scale

Power ratios are often described on a logarithmic scale, especially when the ratio takes on a wide range of values, such as the case of received power divided by transmitted power for the high-loss paths characteristic of echoes from deep regions of the body. The commonly used decibel scale (dB) is defined below:

$$dB = 10 \log_{10}\left(\frac{W_1}{W_2}\right)$$

TABLE 3.1 SUMMARY OF IMPORTANT IMPEDANCE, POWER, AND REFLECTION RELATIONSHIPS

Impedance

$$Z = \frac{p}{u} = \rho_0 c = \sqrt{\frac{\rho_0}{K}}$$

Power Density

$$I_{ave} = \tfrac{1}{2}\frac{p^2}{Z} = \tfrac{1}{2} Z u^2$$

for sinusoidal waves

Reflection at Interfaces

Angles:

$$\frac{\sin \theta_i}{\sin \theta_t} = \frac{\lambda_1}{\lambda_2} = \frac{c_1}{c_2} \quad \text{(Snell's law)}$$

$$\theta_r = \theta_i \quad \text{(Specular reflection)}$$

Magnitudes:

$$R = \frac{p_r}{p_i} = \frac{(Z_2/\cos\theta_t) - (Z_1/\cos\theta_i)}{(Z_2/\cos\theta_t) + (Z_1/\cos\theta_i)} \quad \text{(pressure reflection ratio)}$$

$$\frac{\text{Reflected power density}}{\text{Incident power density}} = R^2$$

$$\frac{\text{Transmitted power density}}{\text{Incident power density}} = (1+R)^2 \frac{Z_1}{Z_2}$$

where dB is the decibel measure of the ratio of power W_1 to power W_2. The decibel scale can just as easily be applied to power density ratios. In addition, if the pressure amplitudes associated with each wave are known and the two waves are in media having the same acoustic impedance, the power decibel measurement can equivalently be given in terms of pressure as

$$dB = 20 \log_{10}\left(\frac{p_1}{p_2}\right)$$

since power density is proportional to pressure squared.

Table 3.1 contains a summary of many of the important equations and relationships derived in this chapter.

PROBLEMS

3.1. Starting with Equation (2.11) of Chapter 2 and using the definition of impedance, Equation (3.1), derive an expression identical to Equation (3.4).

3.2. By using either Equation (3.3) or Equation (3.4) with quantities expressed in correct SI units, show that correct SI units for impedance are kg/m² s. Then, show that these units are consistent with Equation (3.1).

3.3. Referring to the results of Problem 2.8, rank the materials given there in order of increasing acoustical impedance Z.

3.4. Find the *time*-average power density carried by a pressure wave of the form $p = p_+ \cos(\omega t - kz)$ by integrating in time over one cycle (i.e., over one period, $1/f$).

3.5. A typical average power density for an ultrasonic medical imaging instrument is $I_{ave} = 10$ mW/cm². Find the peak pressure and peak particle velocity for such a wave in water, with correct units. Then, using these values, show that $\rho_1 \ll \rho_0$ and also that the approximations which led from Equation (2.5) to Equation (2.6) in Chapter 2 are valid. Use $f = 1$ MHz as a typical frequency.

3.6. Using geometry, show that in order to match the spacing d between projected phase fronts of the incident wave in Figure 3.2, the transmitted wave must have an angle θ_t such that

$$\frac{\sin \theta_i}{\sin \theta_t} = \frac{\lambda_1}{\lambda_2} = \frac{c_1}{c_2}$$

3.7. Derive Snell's law from the simple picture of a marching line of soldiers shown in Figure 3.3 by considering the position of a line at two instances of time, $t = t_0$ and $t = t_0 + \Delta t$.

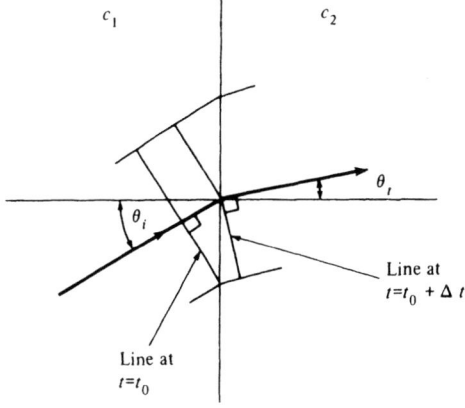

3.8. Snell's law may also be derived by using Fermat's principle, which states that a ray from one fixed point to another will follow a path which minimizes the propagation time between the two points. As an analogy, consider a farmer standing in a flat, dry field who suddenly notices that his tractor is on fire in a freshly plowed neighboring field. The farmer wishes to run to his tractor in the shortest time. His running speed in the dry field is c_1; his running speed in the plowed field is c_2. Using the diagram below, determine the ratio of $\sin \theta_i / \sin \theta_t$ that will minimize his running time to the burning tractor. Compare your answer to Snell's law. (*Hint:* Write the travel time between the two points in terms of L, y, A, B, c_1, and c_2. Then, find the relationship that holds when y has a value that minimizes this time.)

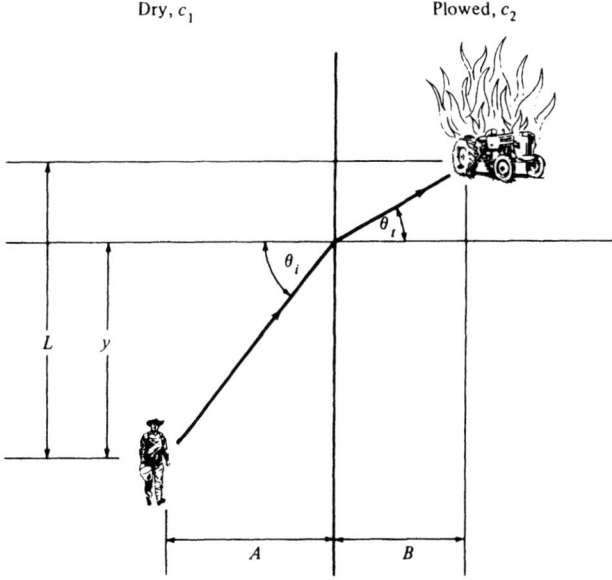

3.9. Find the angle of refraction θ_t for the transmitted ultrasound wave in the configuration shown below:

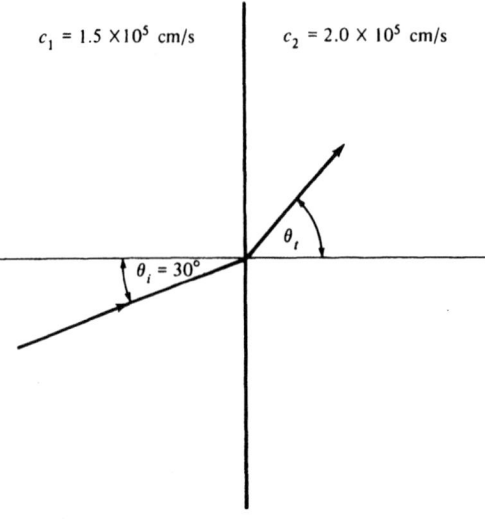

3.10. Derive Equation (3.13) by using the fact that the total pressure on the left side of the interface must equal that on the right side:

$$p_i + p_r = p_t$$

and that the particle velocity components normal to the interface on the left side must equal that on the right side:

$$u_i \cos \theta_i - u_r \cos \theta_r = u_t \cos \theta_t$$

3.11. When passing from a slower region into a faster region, it is possible for a wave to experience Total Internal Reflection (TIR); for certain angles, the entire wave will be reflected, with no transmitted portion. Using Snell's law, solve for the critical angle of incidence θ_c at which TIR begins to occur by setting the transmitted angle equal to $90°$. Then, use this value in Equation (3.13) to find the reflection coefficient R at the critical angle.

3.12. Plot the variation of R with respect to θ_i over the range $-90° \le \theta_i \le 90°$ for the following configuration: In medium 1, $c = 1.5 \times 10^5$ cm/s and $\rho_0 = 1.1$ g/cm^3; in medium 2, $c = 1.2 \times 10^5$ cm/s and $\rho_0 = 1.0$ g/cm^3. Use only enough points in your plot to get the approximate shape of the curve over the range.

3.13. Plot a figure similar to Figure 3.5a except for the case of $Z_2 < Z_1$. Show the correct relationships in magnitude and sign for the incident, reflected, and transmitted waves (both pressure and particle velocity) at the interface.

3.14. Consider the incident pressure wave at normal incidence shown below:
Find the *peak* particle velocity *at the interface*, including units.

PROBLEMS

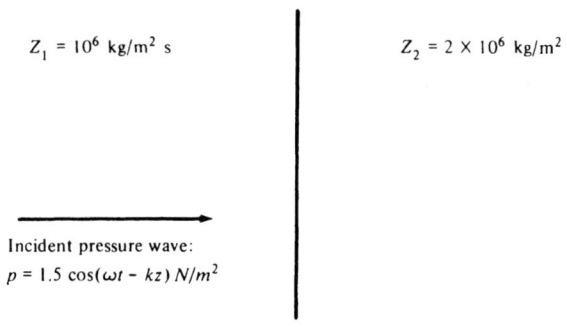

$Z_1 = 10^6$ kg/m² s

$Z_2 = 2 \times 10^6$ kg/m² s

Incident pressure wave:
$p = 1.5 \cos(\omega t - kz)$ N/m²

3.15. A wave with power density of 1 mW/cm² is normally incident on an interface from water into muscle. For muscle, $c \approx 1.57 \times 10^5$ cm/s and $\rho_0 \approx 1.07$ g/cm³.
   (a) Find the power density in the reflected wave.
   (b) Find the power density of the transmitted wave.

3.16. For the results of Problem 3.15, express the ratio of reflected power density to incident power density in dB. Also express the ratio of transmitted power density to incident power density in dB.

3.17. Calculate the reflection coefficient R and the ratio of reflected power to incident power for each of the following cases; assume normal incidence:
   (a) $Z_1 = Z_2$
   (b) $Z_1 = \infty$
   (c) $Z_1 = 0$
   (d) $Z_2 = \infty$
   (e) $Z_2 = 0$

3.18. A wave incident from the left on the boundary shown below has a pressure reflection coefficient given by R_L and a power reflection coefficient given by R_L^2:

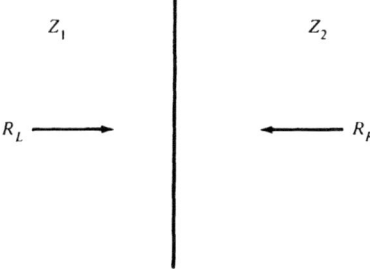

Find the pressure reflection coefficient R_R and the power reflection coefficient for a wave incident on the same boundary from the right. Put both in terms of R_L. Assume normal incidence.

3.19. Prove that the value given in Equation (3.21) is equal to the value given in Equation (3.23), namely, that

$$(1+R)^2 \frac{Z_1}{Z_2} = 1 - R^2$$

only when $\theta_i = 0$. Assume $c_1 \neq c_2$.

Chapter 4
Acoustical Properties of Biological Tissues

4.1 INTRODUCTION

Having developed most of the groundwork for the mathematical treatment of acoustical waves, we now investigate the properties of the media through which these waves propagate in a biological environment. Tissues of various types comprise the terrain probed by the ultrasonic waves, and as might be anticipated from the work of previous chapters, the various impedances, phase velocities, and attenuation rates of the tissues play an important role in determining the effectiveness of the ultrasound to penetrate and image. This chapter begins with a section that surveys the types of tissues found in the body. In the next section the phenomenon of absorption is discussed and, as one possible model for absorption in tissues, the concept of tissue viscosity is introduced. This leads to a loss term in the wave equation. Typical values of acoustic parameters for tissues are then presented.

4.2 SURVEY OF BIOLOGICAL TISSUES

Before proceeding with further wave equation developments, it is instructive to briefly review the general properties of biological tissue types. This survey is not intended to be all-inclusive or to cover anatomy but rather will focus on the tissue characteristics important to understanding ultrasonic propagation.

4.2.1 The Cell

The universal building block of all living tissue is the cell. Cells occur in the body in a tremendous variety of shapes and sizes, although many cells are typically 10 to 100 μm in dimension. Each cell is capable of independent life processes, including reproduction (mitosis), metabolism and oxygen consumption, carbon dioxide production and waste excretion, growth, and repair.

Figure 4.1 shows a schematic drawing of a typical cell and its contents. The cell is bounded by a thin membrane 75 to 100 Å thick which surrounds and packages the contents of the cell. Inside is a gel-like material called the cytoplasm, in which specialized organelles are located to perform the tasks necessary for the proper functioning of the cell. For example, a nucleus with chromosomes important to cell reproduction is found in most cells. Elsewhere in the cytoplasm may be vesicles (with either smooth or rough surfaces) called endoplasmic reticulum whose role it is to partition materials used or produced by the cell in specific chemical processes. Oblong mi-

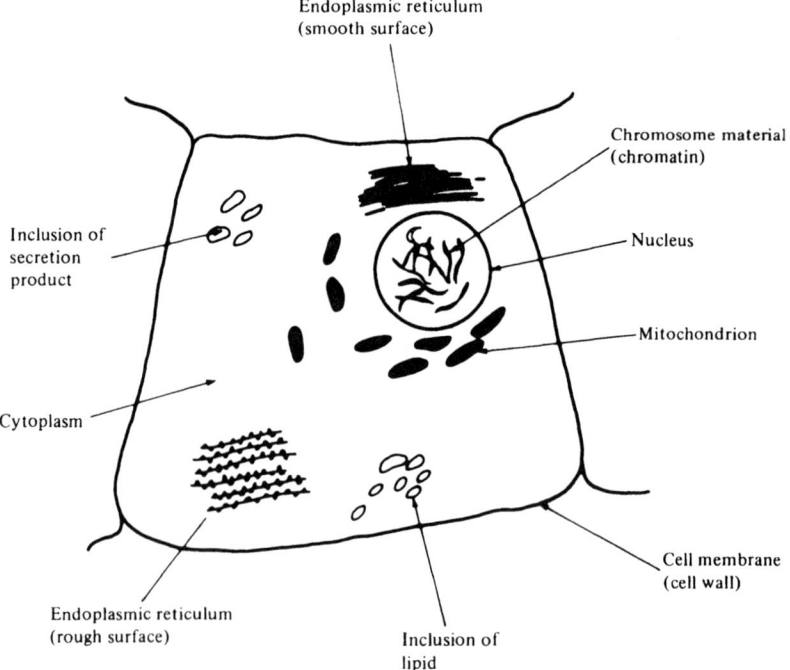

Figure 4.1 Simplified drawing of a typical cell showing some of the organelles important to the functioning of the cell and some inclusions of various materials in the cytoplasm. A membrane encloses the cytoplasm and cell contents.

tochondria are also seen in many cells, especially those of energy-related tissue, since the function of the mitochondria is to metabolize food substances by oxidation, yielding energy for the cell sustenance. In addition, other organelles useful in reproduction, transport, and secretion sometimes are found, along with inclusions or bits of waste material, food, enzymes, lipids, pigment, filaments, and so on.

The effect of ultrasound on the cell depends upon the magnitude of the power density in the exposure beam. At low power levels used in diagnostic imaging, little mechanical trauma or temperature change is experienced by the cell. At very high power levels, however, the cell wall may be damaged or even ruptured, releasing the cell contents and destroying the cell. Also, internal damages to the cell organelles may occur. If many cells are killed by such exposure, the tissue may not be able to repair itself sufficiently, and severe tissue damage may occur. This situation, of course, must be avoided. Chapter 8 of this text gives the exposure limits currently considered safe for tissue irradiation under various exposure conditions.

4.2.2 Tissue Types

The cells are organized into groups that constitute the various tissues of the body. These tissues may be classified into four basic tissue types according to their function: epithelial, muscular, nervous, and connective tissue.

Epithelial Tissue

Epithelial tissue is comprised of cells that form sheets or layers to cover various surfaces of the body. The fundamental role of epithelial tissue is the protection, compartmentalization, and regulation of secretion and intake of substances by the organ it covers. This role determines the shape of the cells in epithelial tissue—basically tightly spaced cells in one or more layers of various heights. Figure 4.2 diagrams some configurations found in epithelial tissue, and Table 4.1 gives one common classification scheme.

When only one layer of cells is present, the tissue is classified as *simple.* The height of the cells in the layer is then classified as *squamous, cuboidal,* or *columnar,* in order of increasing height. Simple squamous epithelium is found as a lining of blood vessels and kidney ducts, and other places where rapid diffusion through the tissue of vital substances such as oxygen or fluid molecules is important. The thicker simple cuboidal and simple columnar tissues are found in the digestive tract, in organ linings, and in glands where greater lining thickness may be useful but secretion or absorption is still required.

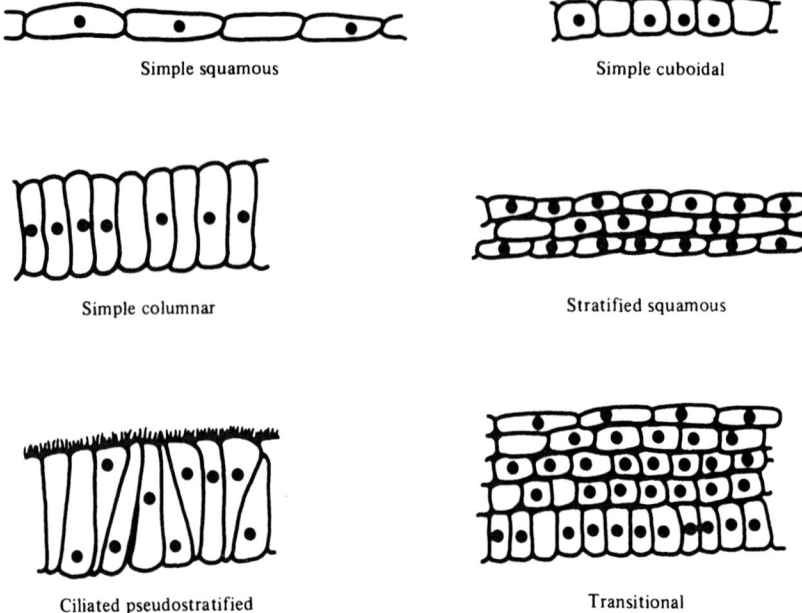

Figure 4.2 Cross section of different classifications of epithelial tissue. Such tissue usually lines cavities, surfaces, or glands.

When more protection or mechanical strength is required, there may be several layers of cells; this tissue is classified as *stratified.* Stratified squamous epithelium is found on the inner and outer coverings of the body (including the skin) where mechanical wear and tear regularly remove cells from the outermost layer.

Much of the mucous membrane lining of the respiratory tract is composed of epithelial tissue where the cells are tall and appear multilayered, but they actually form only a single layer. This tissue is called *pseudostratified* columnar epithelium. Hairlike projections known as cilia are found on the cell's outer surfaces to help transport foreign substance and to increase surface area for secretion or absorption.

TABLE 4.1 CLASSIFICATION SCHEME FOR EPITHELIUM

(Ciliated)	Simple	Squamous	Epithelium
	Stratified	Cuboidal	
	Pseudostratified	Columnar	
		Transitional	

Example: Stratified squamous epithelium

4.2 SURVEY OF BIOLOGICAL TISSUES

Transitional epithelial tissue is a combination of stratified squamous and columnar tissue as shown in Figure 4.2. This allows the tissue to accommodate stretch while maintaining strength, and it is found in such places as the bladder lining.

Being a relatively thin lining, epithelial tissue is normally not a major factor in determining the acoustical properties of most regions of the body. Connective tissue and muscle form a much larger percentage of the volume of most organs and viscera.

Muscular Tissue

Muscular tissue cells are elongated units called fibers, which are usually organized together in bundles of coherently acting fibers. The function of muscle is to provide motion or control of body parts by contraction. Muscle tissue can be classified as *smooth* or *striated,* depending upon its appearance under a light microscope. Smooth muscle is involuntarily controlled and is found in the walls of the digestive tract, in ducts of glands, and in the walls of arteries and veins to control vessel tone. Striated muscle occurs as voluntarily controlled *skeletal* muscle, whose fibers are organized into bundles for locomotion and force generation especially of the limbs, and as *cardiac* muscle, which provides the pumping force of the heart walls.

The microscopically visible striations of striated muscle fibers correspond to regions of repetitive overlap of two kinds of axially oriented long molecules (called the thin and thick filaments) arranged in an alternating fashion down the axis of the fiber. These filaments interdigitate and, upon contraction, slide past one another while producing force. Figure 4.3 is a diagram of some striated muscle fibers. To support the energy requirements, oxygenation, temperature control, and repair of the muscle bundles, many blood capillaries are found in close contact with the muscle fiber bundles.

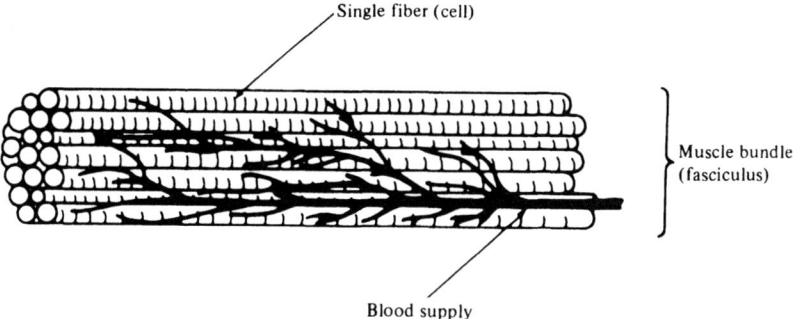

Figure 4.3 Diagram of a section of a striated muscle bundle. Note the tight packing and definite orientation of the fibers and the rich blood supply.

To be effective in force generation, the muscle bundle is generally tightly packed with fibers, and this affects its acoustical properties. The density, phase velocity, impedance, and attenuation values of muscle are all higher than those of water and other loose, soft tissue. These values depend somewhat upon whether the wave propagation direction is oriented parallel or transverse to the muscle fiber's longitudinal axis. For example, attenuation in a direction parallel to the fibers is about twice that encountered when propagating perpendicular to the fibers. However, the orientation effect is usually neglected when there is lack of specific information regarding a muscle's orientation with regard to the beam's direction.

Nervous Tissue

Nerve tissue is used by the body for communication and control. Nerve cells, called neurons, are usually very long cells and act as transmission lines for collecting, transmitting, and distributing nerve impulses. Figure 4.4 shows the parts of a typical neuron. The dendrites project from the end of the cell where the cell body is located and function as multiple receivers or excitation sites in synaptic contact with the endings usually of several other prior neurons. The cell body contains the cell nucleus and cytoplasm. A long projection called the axon carries the nerve's electrical impulses to its end, which may be further differentiated into fine endings where it meets the dendrites of following nerve cells or muscle, glandular, or epithelial cells.

If the neuron is involved in the transmission of impulses over long distances, its axon is generally myelinated, that is, coated with an insulating layer of myelin. Nerve bundles of myelinated axons appear white and form the white matter of the brain and spinal cord. Collections of cell bodies and bundles of unmyelinated axons comprise the gray matter of the brain and spinal cord.

There are an estimated 14 billion nerve cells in man, infusing almost every portion of the body for local control or communication with the brain through the spinal cord. However, only in the brain and spinal cord is the neuron density large enough to affect acoustical propagation. Elsewhere, the nerve bundles are small and spread out, and do not appreciably modify the acoustical properties of the basic tissues (such as muscle) they innervate.

Connective Tissue

This is a broad category of tissue found throughout the body. Connective tissue fills much of the space between organs, providing support and con-

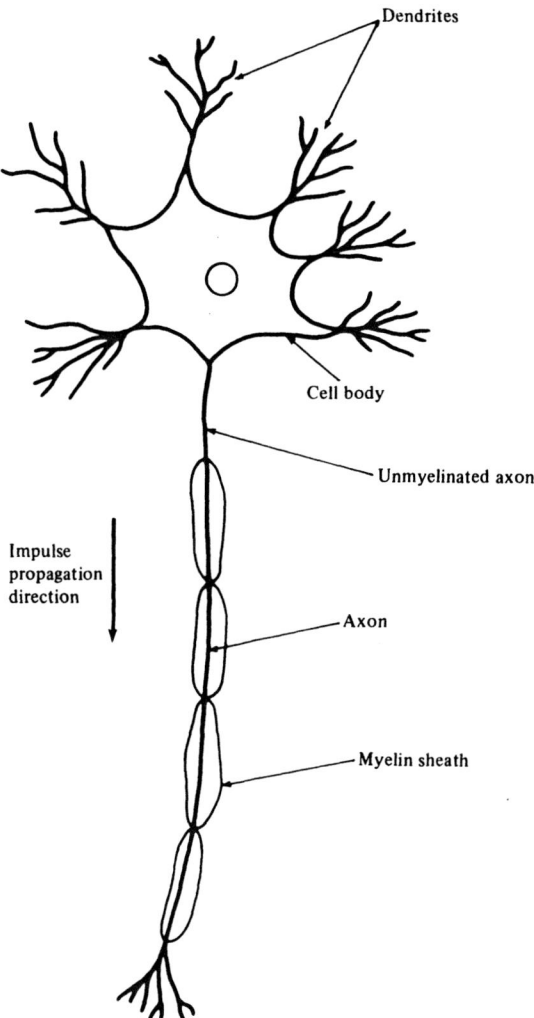

Figure 4.4 Diagram showing the major components of a typical nerve cell (neuron). The myelin sheath electrically insulates the axon of long neurons.

nection to various body parts, as its name suggests. It also may serve as a background material through which cells wander, such as plasma cells for producing antibodies (important to fighting infection) and macrophages (for ridding the body of bits of foreign substances). Due to its ubiquitous presence, connective tissue is found in a vast variety of density, structure, and composition.

One way to classify connective tissue is according to the density and type of nonliving material (usually fibers) found in the tissue's extracellular gel-like ground substance.

1. *Loose* connective tissue is a weblike structure composed of widely spaced fibroblast cells responsible for the generation of the loose and generally unorganized fibers in the tissue. Figure 4.5 shows typical loose connective tissue such as found in the subcutaneous regions of the body. The fibers are not dense, but they still give the tissue its basic mechanical properties. Collagenous fibers have high tensile strength for toughness, while elastic fibers produce resilience in the tissue. Also seen in Figure 4.5 are macrophages and wandering cells for repair and protection of the tissue. Loose connective tissue is often found between organs and filling other anatomical spaces.

2. *Dense* connective tissue is characterized by a greater abundance of fibers in the tissue; these fibers are often organized to provide more strength or elasticity where needed. Tendons, organ capsules, and nerve sheaths are all examples of dense connective tissue.

Two other specialized kinds of connective tissue have important consequences with regard to ultrasonic wave propagation: bone and blood.

3. *Bone* is a type of connective tissue in which the extracellular substance has become hardened through calcification, making the bone strong

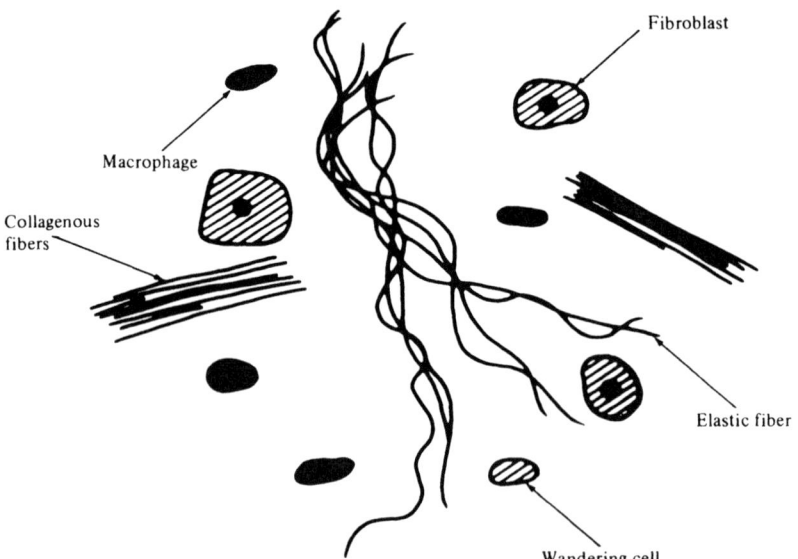

Figure 4.5 A section of loose connective tissue such as found in the subcutaneous region. The density and type of fibers in the extracellular tissue give the tissue its mechanical properties.

and rigid. Its principal role in the body is one of skeletal support, compartmentalization, and protection, as seen in the limbs and skull. Bone also provides a store of calcium to help in calcium regulation in the blood and other fluids, and forms a compartment for bone marrow, which generates red blood cells.

Although the matrix of bone is hardened calcium, bone is a living, adaptable tissue with ample blood supply and cells for generation, protection, and repair. As Figure 4.6 shows, there are numerous canals that infiltrate the bone, carrying metabolites to the bone cells (osteocytes) and containing small blood vessels in the larger canals.

Bone is more dense than soft tissue, but not dramatically so (about 1.7 times) due to the multitude of canals and spaces in the interior of bone. It is, however, much less compressible than soft tissue, which leads to a significantly higher acoustic phase velocity and impedance than in soft tissue. The large mismatch in impedance at a soft tissue/bone interface will result in a high reflection coefficient making it very difficult to penetrate bony areas with ultrasound (see Problem 4.8).

Shear waves will propagate in bone due to the rigidity of its matrix, but these waves are not practical for general soft tissue imaging since they are heavily attenuated by viscosity in soft tissues. Mode conversion to shear waves and scattering contribute to the high absorption coefficient for compressional waves in bone.

4. *Blood* is the fluid which carries nutrients to, and waste products from, all regions of the body. By means of its red blood cells, or erythrocytes, blood transports oxygen from the lungs to all tissues to satisfy their met-

Figure 4.6 Cross section of a bone, showing the many interconnecting canals of various sizes for carrying blood vessels and fluids to the bone cells. Shown are two neighboring Haversian systems, organized around their central Haversian canals.

abolic requirements. In addition, blood can carry heat from or to a region as an important contributor to thermal regulation of the body.

The liquid portion of blood, called plasma, is composed of water with many dissolved electrolytes and protein molecules. An appreciable portion of the total blood volume (about 40%) is occupied by the red blood cells. Also found in smaller concentrations (approximately one per 600 red blood cells) are the white blood cells (leukocytes) of several varieties important to body defenses, and the platelets, which help in the process of blood clotting.

The erythrocytes are highly specialized for the purpose of transporting oxygen. The thin membrane of the erythrocyte surrounds an internal gel containing the iron compound hemoglobin. Hemoglobin has a strong affinity for oxygen at high partial pressure of oxygen, such as in the lung capillaries, where the hemoglobin in the erythrocyte combines with oxygen. At lower oxygen pressures, however, such as in the peripheral tissues of the body, hemoglobin readily releases its oxygen to be used for tissue-metabolic needs.

The shape of the erythrocyte is shown in Figure 4.7. The biconcave disc contour of the membrane is well suited to the function of the cell; its large surface-to-volume ratio facilitates the diffusion of oxygen, carbon dioxide, and other nutrients across the thin membrane to and from the interior region. This shape is also very flexible, and the soft, smooth red

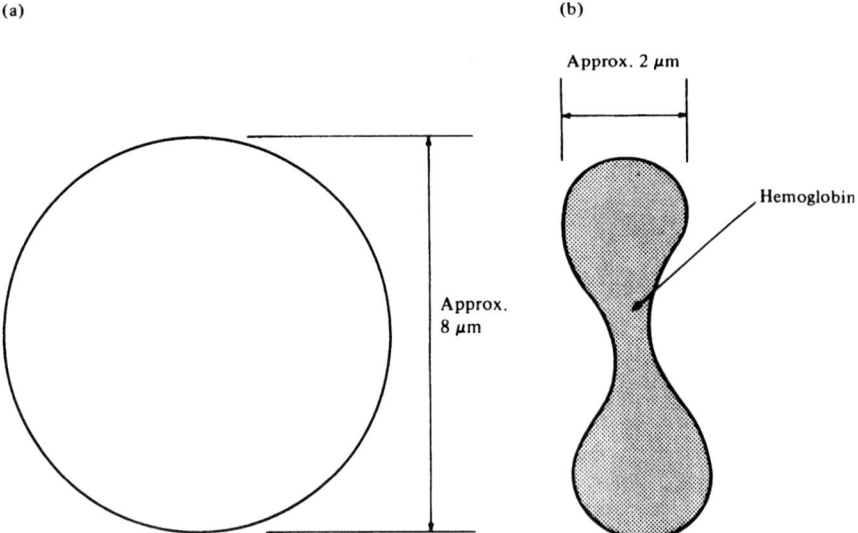

Figure 4.7 (a) A view of a red blood cell looking at its broad surface. (b) A cross section showing the biconcave disc shape of the cell and the internal hemoglobin molecules.

blood cell undergoes much flexing and accommodation as it passes through small capillaries. The mature erythrocyte in mammals has no nucleus and therefore cannot reproduce; new erythrocytes are produced in bone marrow and are continuously fed into the blood stream. The diameter of a human erythrocyte is about 8 μm, and the thickness at the edge is about 2 μm.

Acoustically, the red blood cells in blood plasma act as small point scatterers to an incident ultrasonic wave, thus allowing the measurement of blood flow velocity by the Doppler shift technique (Chapter 7). The acoustic characteristics of the internal material in a red blood cell are not greatly different than those of the surrounding plasma, and the cell membrane is too thin to appreciably affect the wave propagation. Also, since the size of an erythrocyte is much smaller than a wavelength (see Problem 4.2), the scattering by each cell is classified as Rayleigh scattering, generally weak and isotropic. However, the concentration of red blood cells in healthy human blood is so great—about 5×10^6 per mm^3—that the total scattered power from an irradiated volume may be detectable even though each cell scatters only weakly, making Doppler blood flow measurements possible.

4.3 ATTENUATION IN TISSUES

At this point we need to introduce into our wave analysis a term which represents the loss of power that occurs when an ultrasonic wave passes through a length of tissue. The power density of the propagating wave may decrease due to several reasons (see Table 4.2). One is simply that the wavefront may not be planar but rather diverging, leading to a dilution of the wave's energy into an expanding cross-sectional area; this spreading of the beam will be covered in detail in Chapter 5. Another factor that may reduce the beam's intensity is elastic reflection caused by intervening regions of impedance mismatch. If the interfaces are rather distinct and quasi-planar, this can be characterized by the power reflection and transmission coefficients defined in Chapter 3. Even within supposedly homogeneous tissue, however, there may be small, localized variations in acoustic properties which will lead to broad-angle *scattering* of the incident wave and consequent reduction in forward power density. Two examples are red

TABLE 4.2 CAUSES OF WAVE ATTENUATION

Divergence of the wavefront (Chapter 5)
Elastic reflection at planar interfaces (Chapter 3)
Elastic scattering from irregularities or point scatterers (Chapter 7)
Absorption of wave energy (Chapter 4)

blood cell scattering in blood, covered in Chapter 7, and scattering from the multiple air-filled alveoli of lung tissue (where the scattering is so severe that a 1-MHz ultrasound wave is considered nonpenetrating to lung regions).

The major cause of attenuation in most nonlung tissues, however, and the phenomenon we shall deal with in this section, is *absorption* of the wave's energy, which transfers a portion of the originally organized acoustic energy into subsequent heat. The exact cause of absorption by the molecules of biological media is still largely unknown and, in fact, is probably due to a complex variety of interactions. But as a simple model of absorption in tissue which nevertheless sheds some light on the mathematical nature of the energy loss, we shall now introduce the concept of fluid viscosity and show that this property leads to loss.

In Chapters 2 and 3 it was assumed that the fluids supporting the oscillations were completely viscosity free. Equation (2.6) of Chapter 2 showed that the pressure wave could be considered to be the source of a force that caused a time rate of change in the momentum $\rho_0 u$ of the particles in the medium:

$$-\frac{\partial p}{\partial z} = \rho_0 \frac{\partial u}{\partial t} \tag{2.6}$$

In turn, the velocity inherent in this particle momentum caused displacement of the particles from equilibrium, resulting in a restoring force due to the finite compressibility of the elastic material. In this way, the pressure and particle velocity were coupled together, leading to the wave equation. Energy was transferred in the wave as an oscillating combination of kinetic energy due to particle velocity and potential energy due to the elastic force in the pressure wave. No energy was lost because the force was entirely elastic; it could all be recovered to cause momentum change.

In actual fluids, however, not all of the pressure is effective in causing momentum change. Some of the force from the pressure wave must overcome the viscous drag of the particles in the medium as they slide past one another. This resistance is due to the viscosity possessed by the fluid, and it can be modeled as a reduction in the effect of the pressure by introducing an extra viscosity-related term p' in Equation (2.6) which subtracts from p:

$$-\frac{\partial}{\partial z}(p - p') = \rho_0 \frac{\partial u}{\partial t}$$

with

$$p' = \left(\frac{4\eta}{3} + \eta'\right)\frac{\partial u}{\partial z}$$

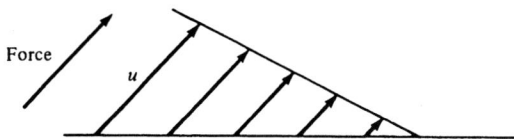

Figure 4.8 Owing to viscosity, a gradient in the particle velocity $\partial u/\partial z$ must be accompanied by a force.

where η is the dynamic coefficient of shear viscosity and η' is the dynamic coefficient of bulk (compressional) viscosity. Substituting for p' yields

$$-\frac{\partial}{\partial z}\left[p - \left(\frac{4\eta}{3} + \eta'\right)\frac{\partial u}{\partial z}\right] = \rho_0 \frac{\partial u}{\partial t} \qquad (4.1)$$

The exact derivation of this extra term p' is left to more advanced texts on the general nature of waves in fluids,* but it is easy to show why the term is proportional to the viscosity coefficients and the velocity gradient $\partial u/\partial z$. Consider a fluid in which the velocity varies as a function of position so that neighboring layers must slide past one another, as shown in Figure 4.8. Due to molecular interaction between layers, a force is necessary to overcome the fluid's resistance to this motion. The magnitude of the force per unit area of interaction will be proportional to the nature of the fluid expressed by the value of its viscosity parameter, and to the degree of sliding expressed by the gradient of the velocity $\partial u/\partial z$. Thus,

$$\left(\frac{\text{Force}}{\text{Area}}\right) \propto \left(\eta \frac{\partial u}{\partial z}\right) \qquad (4.2)$$

A fluid with a relative large viscosity will resist a given velocity gradient with more force than another less viscous fluid (e.g., molasses in January pour more slowly than alcohol). Similarly, for a given fluid, the higher the velocity gradient, the larger the force required.

The extra pressure term in Equation (4.1) is not large in magnitude since η and η' are small numbers for most acoustic media, but its presence leads to an effect not seen before: attenuation of the waves by power loss. The attenuation can be derived by using Equation (2.11) of Chapter 2:

$$\frac{\partial p}{\partial t} + \frac{1}{K} \cdot \frac{\partial u}{\partial z} = 0 \qquad (2.11)$$

Solving for $\partial u/\partial z$ from Equation (2.11) to replace $\partial u/\partial z$ in the new term of Equation (4.1) gives

---

* See, for example P. M. Morse and K. U. Ingard, *Theoretical Acoustics*. New York: McGraw-Hill, 1968.

$$-\frac{\partial}{\partial z}\left[p+\left(\frac{4\eta}{3}+\eta'\right)K\frac{\partial p}{\partial t}\right]=\rho_0\frac{\partial u}{\partial t} \quad (4.3)$$

Combining the partial derivative of Equation (4.3) with respect to z with the partial derivative of Equation (2.11) with respect to t allows the elimination of terms involving the variable u, resulting in a modified wave equation:

$$\frac{\partial^2 p}{\partial z^2}+\left(\frac{4\eta}{3}+\eta'\right)K\frac{\partial^3 p}{\partial z^2 \partial t}-\rho_0 K\frac{\partial^2 p}{\partial t^2}=0 \quad (4.4)$$

A comparison of this equation with Equation (2.14) reveals that an extra term (with three levels of differentiation rather than two) has been added to the wave equation. The consequence of this addition is that all solutions will now have an exponential decay as a function of distance. For example, the typical solution for the pressure wave is now

$$\boxed{p = p_+ e^{-\alpha z} \cos(\omega t - kz)} \quad (4.5)$$

Verification that this is the correct form can be obtained by substituting Equation (4.5) into Equation (4.4) (see Problem 4.3). This procedure leads to an expression for the attenuation constant α:

$$\alpha = \frac{[(4\eta/3)+\eta']\omega^2}{2\rho_0 c^3} \quad (4.6)$$

which is valid for small attenuations where $\alpha^2 \ll k^2$, as is the case in normal tissue (see Problem 4.4). Note that, as expected from its origin in viscosity, the attenuation term is proportional to the viscosity coefficients η and η' of the fluid. Also note that the loss factor (at low frequencies, at least) is proportional to the square of the frequency, ω^2.

Besides attenuation, there are other effects on the propagation characteristics of the waves due to the additional viscosity term. For instance, as shown in Problem 4.3, the phase velocity $c = \omega/k$ is increased slightly due to viscosity. However, for the practical case in tissue where $\alpha^2 \ll k^2$, the change is negligibly small, and

$$c \approx \frac{1}{\sqrt{\rho_0 K}} \quad (4.7)$$

as before. Also, since the force needed to overcome viscosity effects is out of phase from the inertial forces in the system as evidenced by the extra partial derivative in the new term of Equation (4.1), the particle velocity and pressure waves are now no longer exactly in phase. Thus, the impedance value must be changed slightly from a real number to a complex number

4.3 ATTENUATION IN TISSUES

when attenuation is present. Again, however, for the case of small attenuation this change may be neglected so that, as previously,

$$Z \approx \sqrt{\frac{\rho_0}{K}} \qquad (4.8)$$

In most of the practical numerical calculations needed to analyze an ultrasonic bioinstrument, the approximations leading to Equations (4.7) and (4.8) are valid within the accuracy required, and they certainly simplify the calculations of propagation time, reflection coefficient, and angle of transmission. Of course, the attenuation given by Equation (4.5) must always be considered when figuring the amount of power lost in a wave after it has propagated through a given layer of tissue.

It should be mentioned here that there are two other causes of absorption normally discussed when dealing with ultrasound in fluids. One is the loss of power due to the conduction of heat away from regions of higher temperature in the wave toward regions of lower temperature. As a compressional wave propagates through a medium, the density fluctuations of the wave are accompanied by corresponding temperature variations. Within each cycle, heat will diffuse away from that portion of the wave with higher density toward the less dense region. This tends to dissipate the organization of the wave, leading to power loss proportional to the thermal conductivity of the fluid.

The other loss factor is the transfer of energy from the wave into the excitation of molecular vibration levels of the fluid. This effect is more pronounced at some frequencies than at others. In any case, however, both of these absorption causes may be considered as small additions to the viscosity loss modeled in this section for tissues at medical ultrasound frequencies.

Effects of Attenuation

A graphical illustration of how the loss factor introduced into the wave of Equation (4.5) affects its magnitude is given in Figure 4.9. The $e^{-\alpha z}$ term represents an exponential decay in the envelope of the pressure wave's amplitude as a function of distance. The power density of the wave will decrease even faster, since $I = p^2/Z$, so that

$$\begin{aligned} I &= \frac{p_+^2}{Z} e^{-2\alpha z} \cos^2(\omega t - kz) \\ &= I_+ e^{-2\alpha z} \cos^2(\omega t - kz) \end{aligned} \qquad (4.9)$$

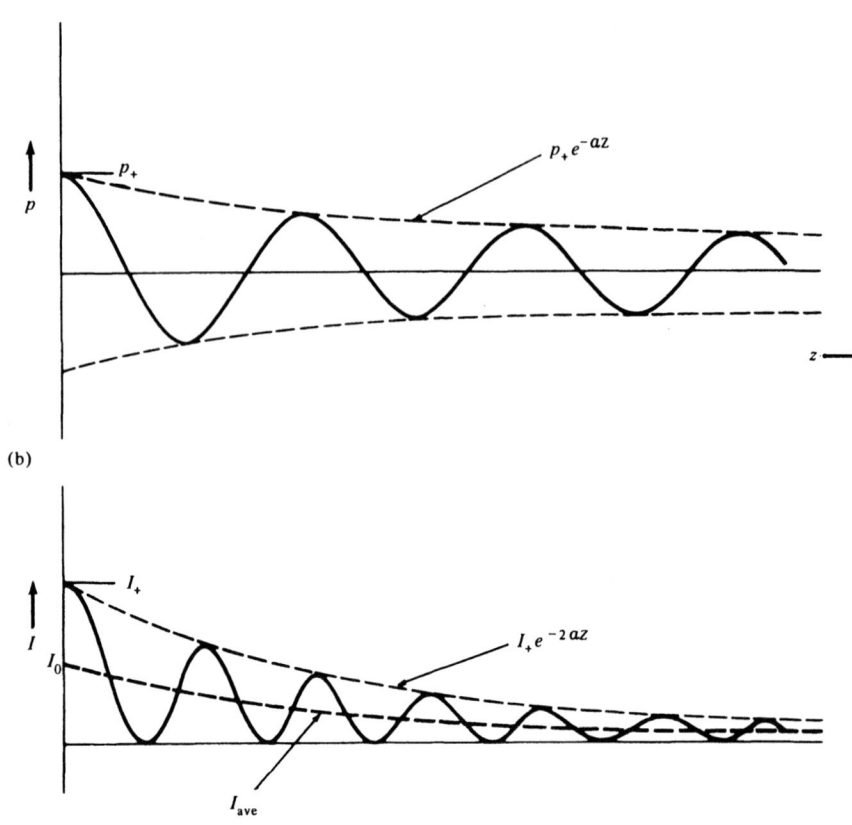

Figure 4.9 (a) Exponential decrease in the amplitude of the pressure wave due to attenuation α; the envelope decreases according to $e^{-\alpha z}$. (b) The power density decreases at twice this rate, $e^{-2\alpha z}$, since $I = p^2/Z$. For the purpose of illustration, the attenuation rate has been greatly exaggerated here. In most tissues, $\alpha \ll k$, so there is only a very small decrease per wavelength.

and the power exponentially decays at a rate of 2α with distance. Since average power density as a function of z is just one-half of the peak value of Equation (4.9), the average power density will also exponentially decay at a rate of 2α:

$$I_{ave} = \frac{I_+}{2} e^{-2\alpha z} = I_0 e^{-2\alpha z} \qquad (4.10)$$

The decrease of amplitude shown in Figure 4.9 has been exaggerated for illustrative purposes. In reality for most tissues, the attenuation factor

α is small compared to the propagation constant k, so $\alpha \ll k$ (Problem 4.4). Since $k = 2\pi/\lambda$, then $\alpha\lambda \ll 2\pi$, and there is little decay *per wavelength*. However, there may still be appreciable loss over the propagation distances encountered in the body. This is because the wavelength is so small in tissue at ultrasound frequencies (on the order of 0.5 mm) that there are hundreds of wavelengths involved in a path of 10 cm.

As a numerical example of attenuation from viscosity, consider water at 35°C. From Chapter 2, $\rho_0 = 1$ g/cm^3 and $c = 1.5 \times 10^5$ cm/s. From experimental data, $\eta = 0.7 \times 10^{-3}$ Ns/m^2 and $\eta' = 2.0 \times 10^{-3}$ Ns/m^2. Thus, Equation (4.6) at a frequency of $f = 1$ MHz gives

$$\alpha_{\text{water}} = 1.7 \times 10^{-4} \quad \text{cm}^{-1}$$

This represents a low attenuation coefficient, and the ultrasonic wave would propagate for several meters in water (degassed to avoid scattering by small bubbles) without losing more than a few percent of its power. Unfortunately, the attenuation values for biological tissues are much higher, as discussed in Section 4.5.

4.4 VISCOSITY RELAXATION IN TISSUES

Actual biological tissues are not ideal viscous fluids, of course, so it is not surprising that the absorption equations of the previous section do not strictly apply. For example, Equation (4.6) predicts that absorption should be proportional to the square of the wave's frequency, while laboratory measurements performed on tissues show that, in the range of medical ultrasound frequencies at least, the actual attenuation is more nearly a linear function of frequency for many soft tissues. The viscosity model is still qualitatively useful, however, after it is corrected for the effect of a finite frequency response of viscosity.

Viscosity is one of a class of phenomena known as *relaxation* effects. An important characteristic of relaxation effects is that they describe behavior that cannot occur instantaneously in the material. Therefore, each effect has an associated time constant (relaxation time) and corresponding relaxation frequency. If we model viscosity as a relaxation phenomenon, its magnitude will drop off as frequency is increased. This means that the viscosity terms in Equation (4.6) are not constant but will decrease with frequency and partially cancel out the increase in the ω^2 term at ultrasound frequencies. The result is a smaller increase with frequency than a power of 2.

A mathematical representation of this viscosity relaxation may be obtained by defining a relaxation frequency ω_i near and above which the frequency response becomes limited. Effective viscosity η_{eff} can then be written as

$$\eta_{\text{eff}} = \frac{\eta_0}{1 + (\omega/\omega_i)^2} \qquad (4.11)$$

where η_0 is the low-frequency viscosity and ω_i is the frequency at which the effective viscosity has been reduced to one-half its low frequency value. Substituting Equation (4.11) into Equation (4.6) for η and η', respectively, each with its own ω_i, shows that the rate of increase of α with frequency will slow down significantly at the higher frequencies.

Actually, there are probably several different relaxation phenomena causing absorption in biological tissue, each with its own unique relaxation frequency. For example, molecular excitation, slight chemical changes, rearrangement of structural disorder, and temperature changes can all accompany the propagation of a wave through tissue (the results of these effects are often included as part of the bulk viscosity). The distribution of the magnitudes and relaxation frequencies for these contributions produce a broadened frequency behavior, with attenuation rising at a rate of less than ω^2 over a broad range of frequencies.

In the ultrasonic range of frequencies, many biological tissues exhibit a variation of attenuation that is nearly *linear* with frequency, except for bone and blood (whose attenuation dependence on frequency lies somewhere between ω^1 and ω^2) and for lung tissue. As a general rule of thumb, if the frequency is doubled, the attenuation coefficient will be doubled. The power loss is then exponentially related to the attenuation coefficient through Equation (4.10).

4.5 VALUES OF ACOUSTIC PARAMETERS FOR BIOLOGICAL TISSUE

The measurement of the acoustic parameters for various biological tissues has only been partially completed. The task is complicated by the fact that there is a great deal of variability from sample to sample in tissue preparations, leading to a variability in the measured values. For example, the acoustic characteristics of fat tissue depend upon the species of animal involved, the region of the body from which the sample was obtained, the homogeneity of the tissue, the temperature of the samples, and the particular experimental technique of the investigator, among other things. Also, whether the preparations were alive, perfused, or fixed before the experiment will modify the values. Further, it is evident that the loss mechanisms in real tissues are more complicated than the viscosity phenomenon presented in the previous sections. Because of these variables, often the best that can be achieved is a range of values representative of a certain class of tissue.

In spite of the variability, it is helpful to catalog the acoustic parameters of common tissues, as presented in Table 4.3. Here, the typical density,

TABLE 4.3 TYPICAL VALUES OF ACOUSTIC PARAMETERS FOR SELECTED HUMAN TISSUES

Tissue type and preparation	Density ρ_0 (g/cm^3)	Phase velocity c (m/s)	Attenuation α at f = 1 MHz (cm^{-1})	Remarks
Whole blood, fresh, heparinized	1.055	1580	0.034 at f = 2MHz	Attenuation dependence approx. $f^{1.25}$
Bone, skull	1.738	2770 ± 185	1.5	Attenuation dependence approx. $f^{1.7}$
Brain, fresh	1.03	1460	0.06	Pathology-free
Breast, in vivo or fresh	—	1510 ± 5	0.22	Pathology-free
Fat, fresh or refrigerated	0.937 (pig)	1479	0.07 ± 0.02	Measurements at 37°C
Heart muscle (beef)	1.048 ± 0.0036	1546 ± 4.7	0.185	—
Kidney (beef)	1.040 (pig)	1572	0.09	—
Liver, fresh	1.064 (pig)	1569.5 ± 4	0.149	Pathology-free
Lung, fresh (dog)	0.4	658	4.3	Inflated; attenuation dependence $f^{0.6}$
Muscle, striated	1.07 (pig)	1566	0.15	Attenuation perpendicular to fibers
Water	1.0	1500	—	Approx. values

Tissue data selected from S. A. Goss, R. L. Johnston, and F. Dunn, "Comprehensive Compilation of Empirical Ultrasonic Properties of Mammalian Tissues," *J. Acoust. Soc. Am.* 64 (Aug. 1978), 423–457; and *J. Acoust. Soc. Am.* 68 (July 1980), 93–108.

phase velocity, and attenuation constant for compressional waves are listed for various materials. In some references the attenuation coefficient is given in the logarithmic decibel scale. To convert to decibel units, the following equation can be used (see Problem 4.5):

$$dB = 8.686\alpha \tag{4.12}$$

where α is the value given in Table 4.3.

It is important to note that the attenuation factor given is the combination of both absorption *and* point scattering effects; thus, lung tissue with its alveolar scattering sites has a very high attenuation. Also, the attenuation factor is often given only for a frequency of $f = 1$ MHz and must be appropriately adjusted for other frequencies. A graph of the variation of α with frequency is shown in Figure 4.10 for some tissue types. Actually, the quantity plotted on the vertical axis in Figure 4.10 is the

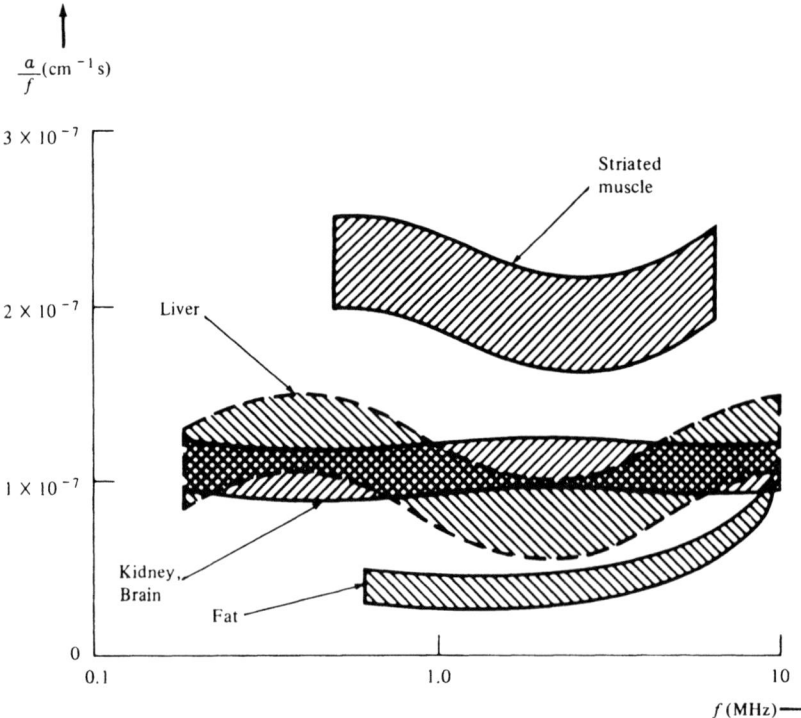

Figure 4.10 Attenuation coefficient normalized to frequency as a function of frequency for several mammalian tissues. Note the large variation within each tissue type at any frequency. Data from D. E. Goldman and T. F. Hueter, "Tabular Data of the Velocity and Absorption of High-Frequency Sound in Mammalian Tissues," *J. Acoust. Soc. Am.* 28 (January 1956), 35–37, Figure 1.

4.5 VALUES OF ACOUSTIC PARAMETERS FOR BIOLOGICAL TISSUE

parameter α/f, which is attenuation normalized to frequency. Two items are evident upon examination of this plot. First, the parameter α/f is reasonably constant over nearly a 100-fold variation in frequency for each tissue type. Thus, the attenuation constant α can be considered to be approximately linear as a function of frequency (except for blood, bone, and lung; see Table 4.3), in agreement with the statements of the previous section. This increasing attenuation sets a practical upper limit to the frequency used in biological imaging. Second, the variability within tissue type is evidenced in Figure 4.10 by bands of values characteristic of each tissue.

Another illuminating way of plotting the acoustic parameters for the materials of Table 4.3 is shown in Figure 4.11. Here, the density and phase velocity are plotted to provide a comparison between the relative magni-

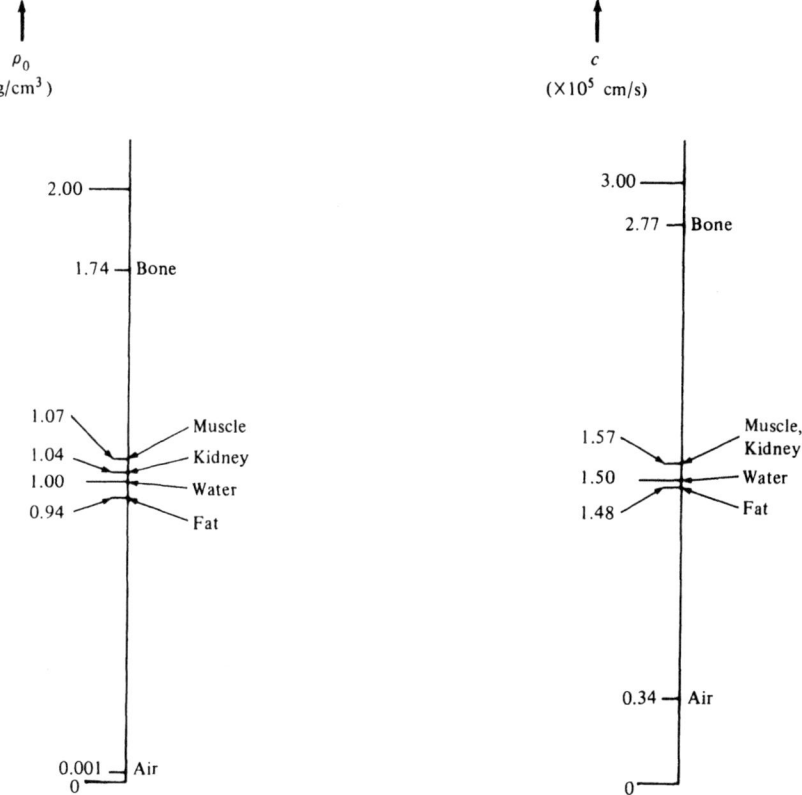

Figure 4.11 Chart showing the relative magnitudes of density ρ_0 and phase velocity c for some common materials and tissues. Since $Z = \rho_0 c$, impedance can be quickly calculated from the above values. Note that the impedances of most soft tissues match well, except those for bone and air.

tudes for the different materials. Note that the soft tissues (muscle, fat, kidney) have values for both parameters which lie very close to that of water. They are clustered within a range of about ±5%. Air and bone, however, are far different in both density and phase velocity from the soft tissue group. This is due to the unique density and compressibility characteristics of air and bone. Air is very light but also can be compressed easily. Thus, ρ_0 is low but K is high, and $c = 1/\sqrt{\rho_0 K}$ is low. Bone is slightly denser than soft tissue but has a low compressibility. Thus, ρ_0 and c are high for bone.

Figure 4.11 is arranged so that tissue impedance can be readily calculated. Since $Z = \rho_0 c$, impedance is found by multiplying the numbers of the left-hand column by numbers from the right-hand column. It is seen that the impedances of most soft tissues match each other (and water) very well, within ±10%. The exceptions are air (much lower impedance) and bone (much higher impedance), which explains the reason why ultrasound will not transmit appreciable power across soft tissue/air or soft tissue/bone interfaces such as are found at the skull, ribs, lungs, or in gas-filled intestinal sections. Such regions block the passage of ultrasound because of the large impedance mismatch and large amplitude reflection coefficient R, and they must be imaged from alternative directions.

From an instrumentation design point of view, the question may be asked: Is it fortunate or unfortunate that most soft tissues have impedance values which are close in magnitude to each other? Since impedance mismatch at an interface causes reflection in proportion to the amount of the mismatch and since reflected power defines the positions of tissue boundaries in most imaging instruments, it seems that a large impedance mismatch would be desirable for maximum resolution of the borders. Yet the penetrating wave must often pass through several interfaces as it probes complex body structures, and large impedance differences would cause a large amount of power to be spent in reflection, leaving little remaining in the transmitted beam for imaging deeper sites. So, moderate to low values for the impedance differences are likely optimum for imaging complex regions in the body.

PROBLEMS

4.1. Give a specific example of each of the four tissue types, then list the place where the example is found in the body and tell what physiological function it performs.

4.2. (a) Estimate the number of erythrocytes in each cm^3 of blood, given the dimensions shown in Figure 4.7 and the fact that the erythrocytes occupy about 40% of total blood volume. Use a circular cylinder model for the erythrocyte.

(b) Compare the size of an erythrocyte to the wavelength of an ultrasonic wave in blood at $f = 2.25$ MHz.

(c) Using the results of part a, find the total surface area of the erythrocytes in a typical adult male whose total blood volume is 5.5 liters. Compare this area to the surface area of the walls of the room you are now occupying.

4.3. Substitute Equation (4.5) into the modified wave equation, Equation (4.4), to get Equation (4.6) for the attenuation constant in fluids for the case of small attenuation:

$$\alpha = \frac{[(4\eta/3) + \eta']\omega^2}{2\rho_0 c^3} \tag{4.6}$$

Also show that the phase velocity $c = \omega/k$ is only slightly modified for small attenuation, so that

$$c \approx \frac{1}{\sqrt{\rho_0 K}}$$

(*Hint:* Separate the wave equation into two parts, one involving sine terms and the other involving cosine terms. To be true at all points in space and time, each part must be satisfied independently. Then, let $\alpha^2 \ll k^2$ for the case of small attenuation.)

4.4. From Equation (4.6) show that the correct SI units for α are m^{-1}. Verify that this is consistent with the use of α in Equation (4.5). Also show that for muscle, $\alpha^2 \ll k^2$ at typical medical ultrasound imaging frequencies, using the values from Table 4.3.

4.5. Attenuation is sometimes specified in units of dB cm^{-1} MHz^{-1} for tissues that exhibit a linear dependence on frequency. Determine a conversion factor that will convert the α values from Table 4.3 into units of dB cm^{-1} MHz^{-1}, then find the attenuation of power density in fat in those new units. Remember that the decibel is defined for *power* ratios.

4.6. An ultrasound wave with an initial power density of $I_0 = 1$ mW/cm^2 and frequency $f = 2$ MHz travels through 1 cm of fat and 5 cm of muscle. How much power density is left in the wave at the end of the one-way travel? Include the effect of reflection at the fat/muscle interface (assuming $\theta_i = 0$).

4.7. Consider the ultrasound imaging configuration shown below:

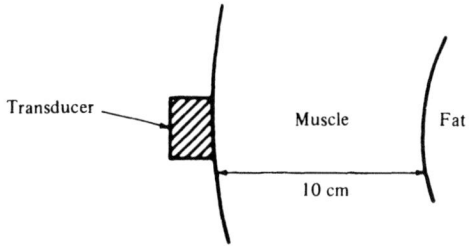

If at least 10 μW/cm^2 must be returned to the transducer from the reflection at the muscle/fat interface for a good signal-to-noise ratio, calculate how much power density must be transmitted by the transducer at a frequency of $f = 1$ MHz. Neglect any losses due to transverse spreading of the beam.

4.8. Calculate the reflection coefficient of power density for each of the following interfaces (assume normal incidence):
   (a) Fat/muscle.
   (b) Brain/muscle.
   (c) Brain/bone.
   (d) Muscle/air.

4.9. In muscle a certain acoustic wave possesses a peak pressure of 35 N/m^2 and has a frequency of 3 MHz. Find the wavelength, the peak particle velocity, and the power density of this wave.

4.10. In the configuration shown below, the average power density produced by the transducer radiating into the fat is 10 mW/cm^2. If the area of the transducer is 10 cm^2 and its frequency is 3 MHz, determine how much average power will be received back at the transducer from the reflection at the fat/kidney interface. Neglect beam spreading.

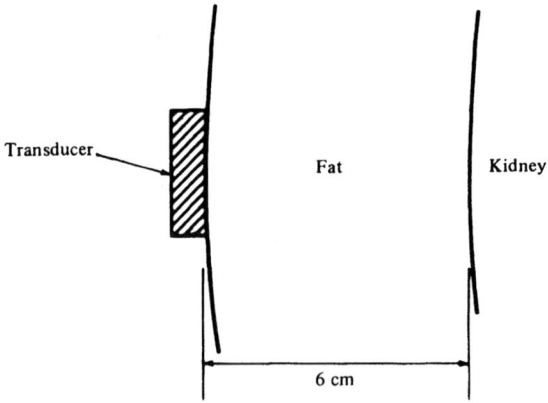

4.11. To investigate the effect of frequency on the attenuation, redo Problem 4.10 for all the same values except assume a frequency of 1 MHz. Compare your answer with that for the original Problem 4.10.

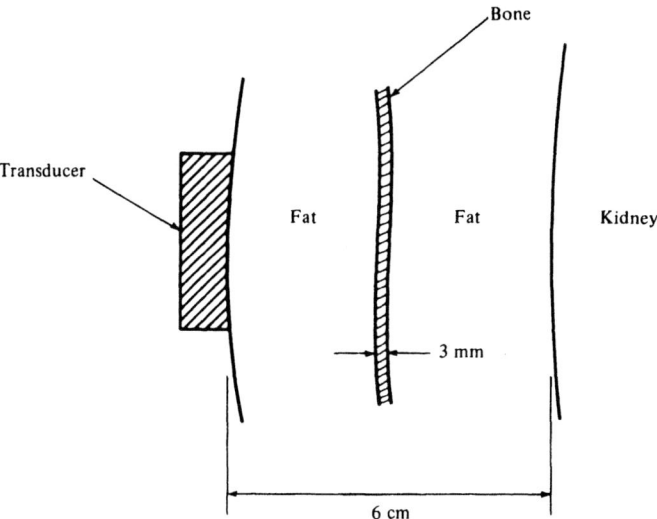

4.12. To investigate the "shadowing" effect of an intervening layer of high-impedance material, redo Problem 4.10 for the same values except assume that a thin layer of bone (3 mm thick) is placed midway in the fat as shown in the figure below. Consider only the primary echo from the fat/kidney interface. Ignore multiple reflections.

Chapter 5

Transducers, Beam Patterns, and Resolution

5.1 INTRODUCTION

A key ingredient of any ultrasonic instrument is the means for generating and detecting the acoustic waves. Since the origin of most generator signals is electrical in nature and since the most convenient way of conditioning, amplifying, and displaying signals is by electronic circuits, some device for translating electrical power into acoustical power, and vice versa, is needed. Among the possibilities are induction coil loudspeakers and magnetostrictive devices, but by far the most convenient transducers at ultrasonic frequencies are piezoelectric crystals and ceramics.

Piezoelectric materials (piezo = pressure) possess the property that a voltage applied to them will produce a pressure field on the atoms in their lattice (a stress) with an accompanying overall contraction or expansion in one or more dimensions of the material (a strain). The stress is a result of the lack of a center of inversion symmetry in the ionic lattice structure of the material; Figure 5.1 shows how an asymmetric atomic structure will distort in an applied electric field. By the piezoelectric property of the material, electrical excitation is changed into motion and pressure, the necessary elements for acoustic waves. Since the process is reversible, a piezoelectric crystal will also change an impinging pressure field into a strain and resulting voltage, so it can be used as an ultrasonic receiver just as well. Certain semicrystalline polymers, such as poly(vinylidene fluoride), PVDF, may also be made piezoelectric by stretching and polarizing them in a strong electric field during fabrication.

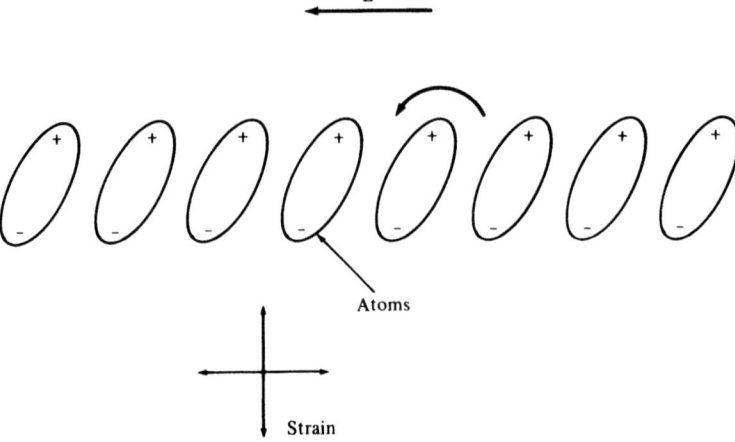

Figure 5.1 When an electric field E is applied to a piezoelectric material, in which charge asymmetry exists on an atomic scale, stresses and strains result in the material. This diagram is highly simplified.

In this chapter we discuss the details of electrical stimulation of piezoelectric transducers, analyze the spatial beam patterns from single transducers, and introduce the concept of multiple-element transducer arrays. It will be shown that the ultimate resolution (lateral and axial) of bioinstruments is determined by the size, frequency, and acoustical "Q" of the transducer used.

5.2 ELECTRICAL EXCITATION OF PIEZOELECTRIC TRANSDUCERS

Figure 5.2a shows a simplified diagram of a piezoelectric material cut and oriented for use as an ultrasonic transducer. The material might be quartz, barium titanate, lead zirconium titanate (PZT), or poly(vinylidene fluoride) (PVDF). Two opposite faces of the transducer are plated with conductive metal films; a voltage generator V is attached to the electrodes to produce an electric field E_z across the thickness l of the transducer whose magnitude is given by (assuming the diameter is much larger than l)

$$E_z = \frac{V}{l} \qquad (5.1)$$

In piezoelectric materials in general, any given orientation of the electric field might produce two stresses (shear and compressional) in any of the three directions of the crystal, so a complete specification of the piezoelectric properties of the crystal would require a 3×6 tensor to tell

(a)

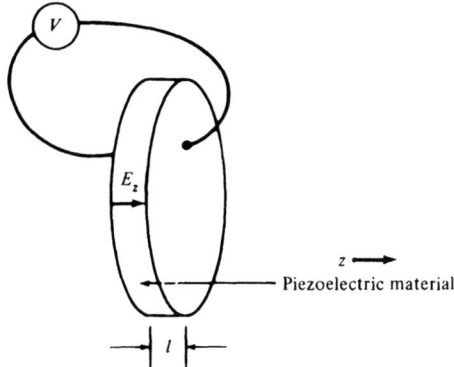

(b)

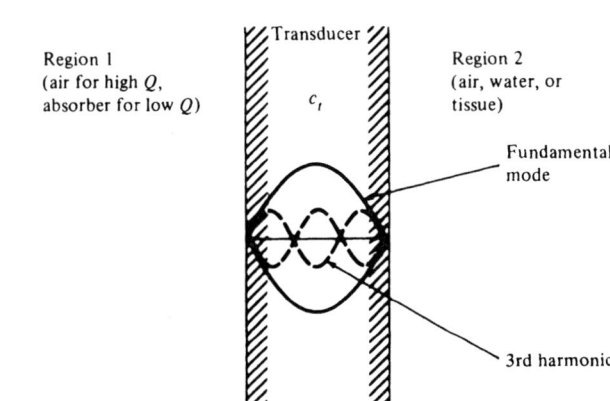

Figure 5.2 (a) Simplified sketch of a piezoelectric material used as a transducer with opposing electrodes. (b) In order to match excitation and boundary conditions, an odd number of half-wavelengths must fit between the transducer faces.

how each of the six components of the stress is related to each of the three components of the electric field. However, in practice the material is usually oriented to take advantage of the largest piezoelectric coefficient, which for most materials is one for which compressional stress is in the same direction as the applied electric field along some preferred axis. For the orientation shown in Figure 5.2a, known as the "thickness" mode of vibration, the pressure on the broad faces will be mainly longitudinal and the resulting pistonlike action will set up the desired compressional waves. The piezoelectric coefficient relating the resultant stress to the electric field in this case is labeled as either e_{11} or e_{33}, depending upon the convention used for the particular crystal or ceramic.

There are two interesting possibilities for the temporal nature of the electrical excitation to the transducer—continuous wave (cw) and pulsed. These two cases are covered in order next.

5.2.1 Continuous Wave Excitation

If the voltage generator applies a voltage across the transducer of the form $V = V_0 \cos \omega t$, then the pressure waves produced will be continuous sinusoidal-type waves of the nature discussed in earlier chapters. These waves will propagate inside the crystal with a phase velocity c_t and will strike the front and back faces of the crystal. Here, they will be reflected in proportion to the impedance mismatch between the crystal material and the materials outside. Since the impedance of the transducer material is generally much higher than that of the air, water, or tissue media against the transducer faces, the reflection coefficient will be nearly $R = -1$, so the resultant pressure at the two boundaries must be nearly zero and a standing wave will be set up inside the transducer between its faces.

Only certain frequencies of excitation will be effective in generating waves that have the proper wavelength inside the transducer to match the simultaneous requirements for zero pressure at both interfaces. These frequencies, called the resonant frequencies of the transducer, are those for which an integral number of half-wavelengths fit between the faces of the transducer cavity. In addition, because the electrical excitation has the same polarity across the entire thickness of the transducer at any given instant (since electrical wavelength is much larger than l), only standing wave patterns with an odd number of half-wavelengths will be efficiently driven by the electrical input. Patterns with an even number of half-wavelengths will always have an equal number of regions with opposite phases, which will cancel electrically, leading to minimal coupling with the input field.

Figure 5.2b shows two waves which match both the boundary conditions and the excitation requirement. The lowest frequency to satisfy the resonance condition is called the fundamental frequency of the crystal, and at this frequency a single half-wavelength fits inside the cavity. The nulls of the pressure standing wave occur at the faces of the transducer to match the boundary conditions. For a transducer of thickness l, the fundamental frequency f_1 will have a wavelength λ_1 inside the transducer such that

$$\frac{\lambda_1}{2} = l \qquad (5.2)$$

5.2 ELECTRICAL EXCITATION OF PIEZOELECTRIC TRANSDUCERS

Since $\lambda_1 = c_t/f_1$, where c_t is the compressional wave velocity in the transducer material, then

$$\boxed{f_1 = \frac{c_t}{2l}} \quad (5.3)$$

At ultrasonic frequencies, the thickness required to use a transducer crystal in its fundamental mode can be quite thin, making some crystals fragile (see Problem 5.1), so very-high-frequency transducers are sometimes employed in their higher harmonic modes. As an example, Figure 5.2b shows a third harmonic wave which will oscillate at three times the frequency of the fundamental.

Frequency Response

Near each of the resonant frequencies, the transducer will have a response to voltage that will vary according to the proximity of its frequency to the resonant frequency. A curve showing how the power density I radiated by a transducer varies as a function of frequency around its point of resonance is given in Figure 5.3. The narrowness or broadness of the resonance curve, as measured by the frequency width Δf to the half-power points, is defined by the so-called quality factor, or Q, of the cavity in the following way:

$$\frac{f_1}{\Delta f} = Q \quad (5.4)$$

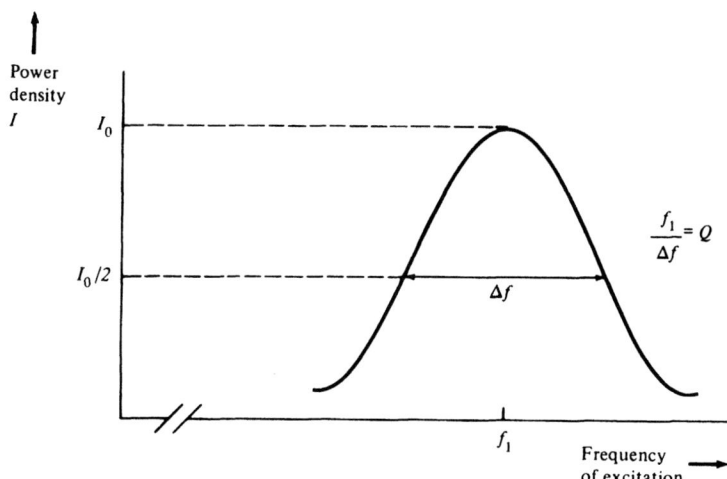

Figure 5.3 The resonance curve for a transducer with center frequency f_1 and quality factor Q. The larger Q, the narrower the frequency response.

Thus, a high Q leads to a very narrowly peaked resonance, and a low-Q transducer has a broadband response.

The magnitude of Q is determined by the losses (absorption and transmission) encountered in the transducer. By far the largest contributor to the losses of most transducers is the transmission of acoustic power through the faces into neighboring regions, since the internal loss of good transducer materials, especially quartz, is small. If air forms the regions on both sides of the transducer, the impedance mismatch is so large that hardly any power escapes, leading to Q values as high as 30,000. In fact, for use in high-precision frequency oscillators, quartz crystals are mounted in small evacuated cans where the vacuum environment gives very low transmission losses with Q values approaching 1,000,000. Since Δf is so small for these crystals, they are in common use in electronic equipment whenever accurate frequencies are needed, as in quartz watches.

Of course, if the transducer is to be used for radiating acoustic waves into tissues, some power is purposely lost through one face of the transducer. When tissue replaces air at one of the transducer faces, the impedance mismatch is reduced, power is transmitted, and the Q of the cavity goes down dramatically.* Problem 5.7 shows that for a typical rigid crystal transducer such as quartz radiating into tissue, $Q \approx 5$-15.

Since air presents a large impedance mismatch with the transducer (as compared to tissue), no air can be allowed to find its way between the transducer face and the tissue surface being irradiated if maximum power transmission into tissue is desired. Any air layer more than a fraction of a wavelength in thickness will reflect considerable power back into the transducer, reducing its effectiveness as a transmitter. Thus, in clinical practice, mineral oil or commercially available gel is used to coat the transducer and force out any air between the transducer/tissue interface.

Radiated Power

The power density that a transducer driven by a voltage source will radiate into a medium may be found by using the piezoelectric relationship for the transducer:

$$p_i = e_{ii} E_i - c_{ii}\left(\frac{\partial \xi}{\partial z}\right) \tag{5.5}$$

where

p_i is the pressure in the transducer material

E_i is the electric field applied

---

* The resonant frequency of the transducer is also slightly lowered from its lossless value, because of the loss now encountered.

5.2 ELECTRICAL EXCITATION OF PIEZOELECTRIC TRANSDUCERS

e_{ii} is the material's piezoelectric stress coefficient

ξ is the displacement of the particles in the material, so $\partial\xi/\partial z$ is strain (elongation or compression) of the material

c_{ii} is the elastic stiffness constant of the material

i is a subscript denoting the directions of the pressure, electric field, and strain (here assumed to be all in the same direction).

The analysis then assumes two countertraveling acoustic waves inside the transducer, as diagrammed in Figure 5.4. When combined, these two waves produce the standing wave pattern described earlier. By matching boundary conditions at the two faces for continuity of both pressure and velocity across the interfaces, similar to the procedure of Section 3.4.2, and using Equation (5.5), it can be shown (see Problem 5.3) that the velocity of the transducer faces at resonance is

$$u_f = \pm \frac{2e_{ii}E_i}{Z_1 + Z_2} \qquad (5.6)$$

where Z_1 and Z_2 are the acoustic impedances of the media on either side of the transducer, and the $\pm$ sign denotes that the face velocities are in opposite directions since a resonant vibration mode with an odd number of half-wavelengths was assumed.

Many transducers have air in the region to the rear; for this case, $Z_1 \approx 0$ and Equation (5.6) gives the velocity of the front face (touching tissue or water) as

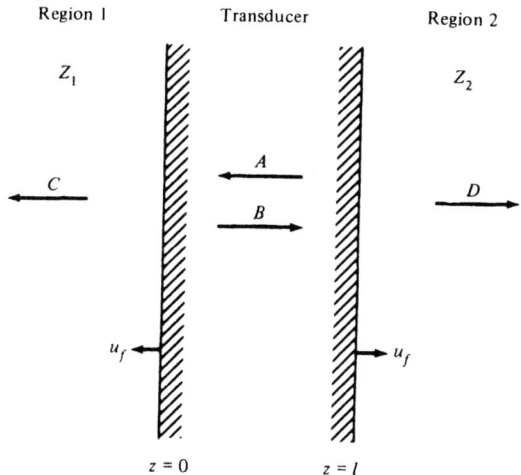

Figure 5.4 Analysis of waves excited inside a transducer. Matching the velocity and pressure boundary conditions at the two faces leads to Equation (5.6).

$$u_f = \frac{2e_{ii}E_i}{Z_2} \qquad (5.7)$$

The power density transmitted forward into medium 2 is then easily found from velocity continuity and the relationship $I = Zu^2$ to be

$$I = \frac{4e_{ii}^2 E_i^2}{Z_2} \qquad (5.8)$$

For the configuration shown in Figure 5.2a where the transducer is excited by a sinusoidal voltage source, Equation (5.1) may be used to give the average radiated power density:

$$\boxed{I_{ave} = \frac{2e_{ii}^2 V_0^2}{l^2 Z_2}} \qquad (5.9)$$

where V_0 is the peak sinusoidal exciting voltage and a factor of $\frac{1}{2}$ was used to give the time average of power density.

Piezoelectric Coefficients

As Equation (5.9) shows, the ability of a transducer to convert voltage into acoustical power is related to the strength of its piezoelectric stress coefficient e_{ii}. Table 5.1 gives some values for various piezoelectric materials commonly used as ultrasound transducers. Sometimes the literature will list other related coefficients, such as d_{ii}, the piezoelectric strain coefficient (sometimes called the transmitting constant). It is related to e_{ii} by the relationship

$$e_{ii} = d_{ii} c_{ii} \qquad (5.10)$$

where c_{ii} is the material's elastic stiffness constant (under conditions of constant electric field). Also, the piezoelectric coefficient g_{ii} (sometimes called the voltage output coefficient, or receiving constant) may be given. It is related to d_{ii} by

$$d_{ii} = g_{ii} \epsilon_r \epsilon_0 \qquad (5.11)$$

where ϵ_r is the relative dielectric constant of the transducer material (under unrestrained or free conditions), and ϵ_0 is the permittivity of free space ($\epsilon_0 = 8.85 \times 10^{-12}$ F/m).

By scanning Table 5.1 it can be seen that there are large differences in e_{ii} and ϵ_r (and therefore in d_{ii} and g_{ii}) among the materials. It would appear that barium titanate or PZT are by far the most efficient radiators, and indeed they are widely used as good transducer materials. But the picture is more complicated than just comparing the values of e_{ii}, since other factors must be considered, such as the electrical coupling of the

TABLE 5.1 VALUES FOR SOME PIEZOELECTRIC TRANSDUCER MATERIALS

Material	Density ρ_0 (kg/m^3)	Elastic stiffness c_{ii} (N/m^2)	Phase velocity c_i (m/s)	Acoustic impedance Z_c (kg/m^2 s)	Relative dielectric constant ϵ_r	Piezoelectric stress coeff. e_{ii} (N/V m)
Quartz (x-cut)	2.7×10^3	86×10^9	5.8×10^3	15×10^6	4.5	.17
Barium titanate	5.7×10^3	110×10^9	5.3×10^3	30×10^6	1700	8.6
Lead zirconium titanate (PZT)	7.5×10^3	83×10^9	4.0×10^3	30×10^6	1200	9.2
Poly(vinylidene fluoride) (PVDF)	1.8×10^3	3×10^9	1.4×10^3	2.5×10^6	12	.069

transducer to the transmitting and receiving circuitry, the internal losses of the material, the material's phase velocity and dielectric constant, the temperature range allowable, and physical attributes such as flexibility and ease of fabrication. For example, for a fixed frequency of resonance, Equation (5.3) shows that a transducer's thickness l is proportional to the material's phase velocity c_t. Therefore, a transducer made from PVDF, because of its relatively low c_t, will be thinner than one of barium titanate. Consequently, since l^2 appears in the denominator of Equation (5.9) for output power, the electric field is high for a given voltage, and PVDF is not as weak as would be predicted by its low value of e_{ii} alone.

Equivalent Circuits of Transducers

Electrical characterization of the transducer is very important in determining the electrical load that the transducer presents to the drive or receiver circuitry and in optimizing the match between the two. As a step in finding the equivalent electrical circuit, a companion equation to Equation (5.5) gives the surface charge density σ_i appearing at the face of the transducer:

$$\sigma_i = \epsilon_r \epsilon_0 E_i - d_{ii} p_i \tag{5.12}$$

This surface charge, occurring on the two parallel electrodes separated by the thin piezoelectric material, forms the essence of a parallel-plate capacitor. The capacitance C_0 of such a parallel-plate capacitor is given by the well-known equation

$$C_0 = \frac{q}{V} = \frac{\sigma_i A}{E_i l} = \frac{\epsilon' A}{l} \tag{5.13}$$

where

q = total charge on either plate
V = voltage between plates
A = area of plate (transducer area)
l = spacing between plates
ϵ' = effective dielectric constant of material between plates.

Due to the piezoelectric activity of the transducer material represented by the second term on the right-hand side of Equation (5.12), the effective dielectric constant ϵ' of the material when used in a nonfree ($p \neq 0$) condition is different from its free value of $\epsilon_r \epsilon_0$. To determine ϵ' for an important nonfree situation, namely, when the transducer is clamped so that all strain is zero, Eq. (5.5) with $\partial \xi / \partial z = 0$ is substituted into Equation (5.12) for p_i:

$$\sigma_i = \epsilon_r \epsilon_0 E_i - d_{ii} e_{ii} E_i = \epsilon_r \epsilon_0 \left(1 - \frac{d_{ii} e_{ii}}{\epsilon_0 \epsilon_r}\right) E_i$$
$$= \epsilon_r \epsilon_0 (1 - g_{ii} e_{ii}) E_i$$
$$= \epsilon_r \epsilon_0 (1 - \kappa^2) E_i \tag{5.14}$$

5.2 ELECTRICAL EXCITATION OF PIEZOELECTRIC TRANSDUCERS

where $\kappa = \sqrt{g_{ii}e_{ii}}$ is known as the coefficient of electromechanical coupling of the material. It can be shown that the parameter κ is related to the ratio of mechanical energy to electrical energy stored in the vibrating transducer.

Using Equation (5.14) in Equation (5.13) gives the material's effective dielectric constant ϵ' under clamped conditions:

$$\epsilon' = \epsilon_r \epsilon_0 (1 - \kappa^2) \tag{5.15}$$

and the capacitance of the transducer when clamped:

$$\boxed{C_0 = \epsilon_r \epsilon_0 (1 - \kappa^2) \frac{A}{l}} \tag{5.16}$$

A detailed electrical analysis of a transducer at resonance (see Problem 5.4) shows that the transducer appears electrically to be composed of just two elements: a capacitor of value C_0 given by Equation (5.16), which represents the accumulation of surface charge on the plates due to the applied voltage; and a parallel resistor R_m, which represents the transformation of electrical power into radiated acoustical power. The value of R_m, called the *motional resistance*, can easily be found from Equation (5.9) and the fact that, for this transducer with no assumed internal losses, all average electrical power $V_0^2/2R_m$ consumed must equal the average acoustical power $I_{ave}A$ radiated. Using Equation (5.9) in this equality,

$$\frac{2e_{ii}^2 V_0^2 A}{l^2 Z_2} = \frac{V_0^2}{2R_m}$$

or, solving for R_m,

$$\boxed{R_m = \frac{l^2 Z_2}{4e_{ii}^2 A}} \tag{5.17}$$

Figure 5.5a shows the equivalent electrical circuit at resonance, with the values of C_0 and R_m given by Equations (5.16) and (5.17), respectively. The capacitance C_0 can be moderately high (due to the large values of ϵ_r for many transducer materials, as large as $\epsilon_r = 1700$ for barium titanate), and the resistance R_m is inversely proportional to the power radiated by the device; high acoustic radiating ability means a low value for the parallel R_m, and vice versa. Problem 5.5 gives values typical of a medical imaging transducer.

As the circuit of Figure 5.5a shows, a transducer appears capacitive in nature right at its frequency of resonance. So, to efficiently match it to the driving voltage generator, a parallel inductor L_0 is sometimes placed between the transducer and generator. The value of the inductor is chosen such that the electrical resonance frequency $\omega = \sqrt{1/L_0 C_0}$ is matched to the acoustical resonance frequency. A transformer may also be used to

(a)

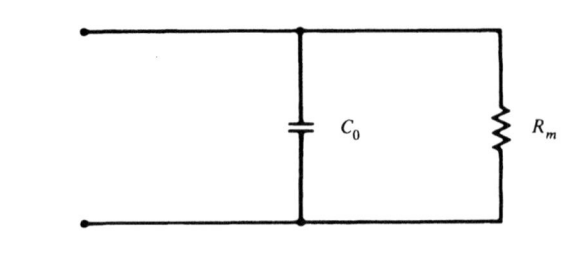

(b)

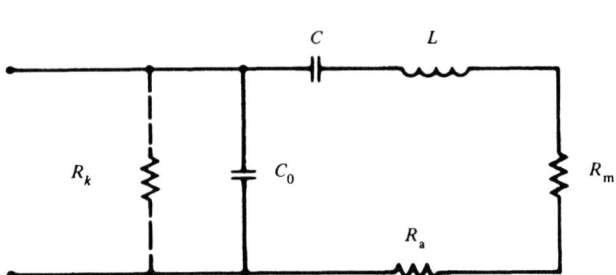

Figure 5.5 (a) The equivalent electrical circuit for a lossless transducer precisely at resonance. C_0 is the parallel-plate capacitance of the device (with a modified dielectric constant ϵ' to account for the piezoelectric activity), and R_m is a resistance representing the radiation of acoustic power. (b) The complete equivalent circuit in the neighborhood of resonance adds an inductance L and capacitance C in series with R_m. To complete the picture, possible internal loss resistances R_k and R_a may also be added. At resonance, the impedances of L and C cancel.

transform the transducer's resistance at resonance, R_m, to match the output impedance of the generator, usually 50 ohms. However, the addition of an electrical resonance circuit increases the overall electrical Q of the network, and in some applications, such as when short acoustic pulses are required for echo ranging, a high Q is not desirable; the effect of Q on pulse length is covered in the next section.

When the frequency driving the transducer is moved away from resonance, two more components are needed to characterize the equivalent circuit: an inductor L and a capacitor C in series with R_m. The impedance of these two elements cancel right at resonance but give this branch a capacitive nature below resonance and an inductive nature above resonance.

To complete the picture, two more resistors may be added to account for any nonradiative losses in the transducer: a parallel resistor R_k (generally

5.2 ELECTRICAL EXCITATION OF PIEZOELECTRIC TRANSDUCERS

large) to account for leakage current, and a series resistor R_a (generally small) to account for internal absorption in the material. Figure 5.5b shows the complete equivalent circuit, which is valid in the neighborhood of resonance as well as at resonance.

Comparison of Piezoelectric Materials

Returning now to Table 5.1, it can be noted that the top three materials listed in the table are fairly dense, rigid crystals. Quartz occurs both as a natural crystal or may be man-made (SiO_2). Barium titanate and PZT are man-made ceramics that are rendered piezoelectric by first heating above their Curie temperature, then cooling in the presence of a strong electric field to produce a permanent "ferroelectric" effect. These man-made materials may be molded during fabrication to the desired diameter and thickness; sometimes a concave face is molded into the tissue side of the transducer to give focusing of the radiated beam. Note that these three crystals have high acoustic impedances (compared to soft tissue impedance of about 1.5×10^6 kg/m^2 s) due to their dense and relatively incompressible nature.

The polymer transducer material PVDF is much softer and less dense than the other materials. As such, it may be fabricated as a film and has the possibility of being shaped around nonplanar body surfaces. It is fabricated by first stretching the raw material along one direction, then polarizing it in a strong dc electric field. The acoustic impedance of PVDF is a much closer match to that of tissue, and therefore more power is coupled out into the tissue. This lowers the Q of the transducer (see Problem 5.9), making it more broadband and giving it better axial resolution, as discussed in Section 5.2.2. Unfortunately, these advantages are offset somewhat by the larger internal loss that PVDF has compared to the crystalline or ceramic materials, by its lower temperature range of operation (restricted to below about 80°C for continuous exposure, which limits the amount of power it can handle as a transmitter due to heat generation by its internal loss), and by its generally lower piezoelectric transmission coefficients e_{ii} and d_{ii}.

When used in the receiver mode, though, the concern is not so much with the efficiency of the transducer in transforming electrical energy into acoustical energy. Rather, the receiver element is often connected to a high-input impedance voltage amplifier, and a good measure of receiving sensitivity is the voltage output coefficient $g_{ii} = d_{ii}/\epsilon_r \epsilon_0$. Due to the low relative dielectric constant of PVDF ($\epsilon_r = 12$), its voltage output coefficient is high, making it a better receiving material than an efficient energy transmitting element.

We now turn our attention to the other basic way of exciting ultrasonic transducers—with a sharp pulse of electrical voltage. This mode of operation is actually the most common for medical instrumentation, inasmuch as the majority of these imagers use pulsed echoes to locate and image the deep-lying tissue boundaries within the body. The precision with which the boundaries are located along the direction of the beam travel (axial resolution) will be shown to be directly related to the time behavior of the transducer's response to the input voltage pulse, as characterized by the Q value of the transducer.

5.2.2 Pulsed Excitation and Axial Resolution

If the electrical input to the transducer is a sharp impulse of voltage, such as that obtained by rapidly discharging a capacitor using a circuit similar to that shown in Figure 5.6, the pressure wave radiated by the transducer will take the form of an exponentially decaying sinusoid. The voltage pulse may be either negative or positive with respect to ground; a negative pulse is often easier to generate with a positive supply using the circuit shown in Figure 5.6.

Figure 5.7 shows the example of a positive voltage pulse and the resultant pressure waveform from the transducer. The pressure waveform does not precisely duplicate the waveform of the voltage (i.e., a sharp pulse of pressure) because the crystal possesses resonant qualities as discussed in the previous section. When excited by an impulse, the crystal will resonate sinusoidally at its fundamental frequency; the envelope of this wave will decay at a rate proportional to the losses (internal and transmitted) of the

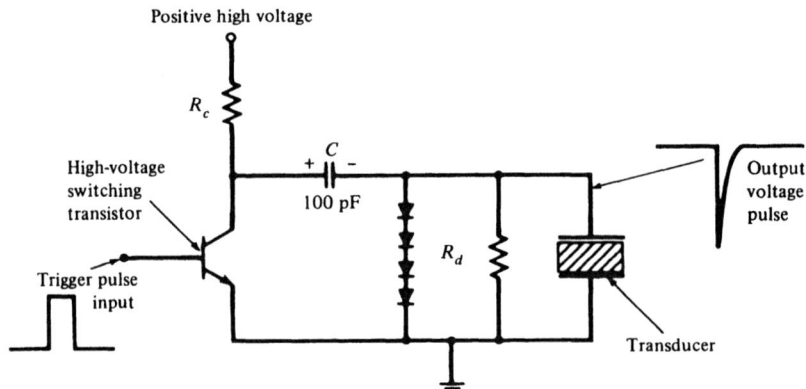

Figure 5.6 An electrical circuit for generating a sharp voltage pulse to a transducer. During the off-time of the transistor, the capacitor charges to the high supply voltage. When the transistor is turned on by the trigger pulse, its low on-resistance takes the left side of the capacitor to near ground voltage, applying a large negative pulse to the upper transducer terminal. The capacitor then discharges through the transducer. R_d is a damping resistor for shaping the trailing edge of the pulse.

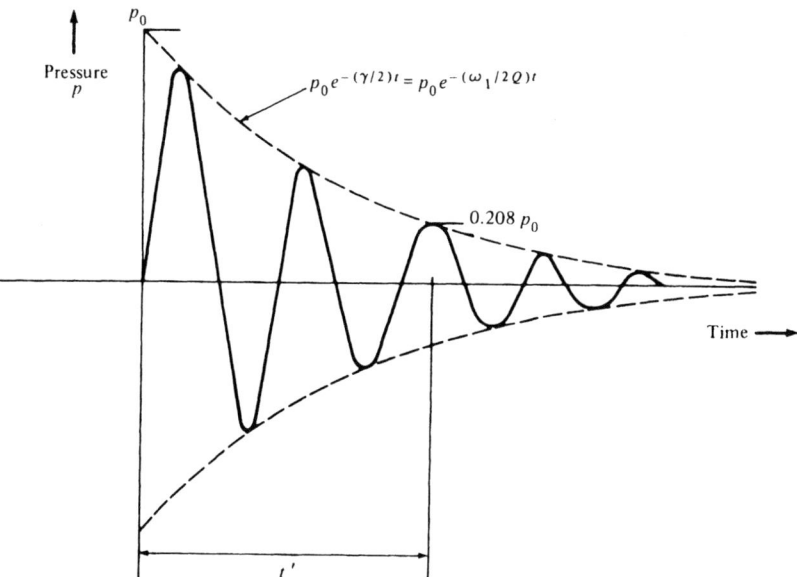

Figure 5.7 The pressure waveform radiated by a transducer excited by a sharp impulse of voltage. The pressure at any distance decays at a rate inversely proportional to the Q of the transducer. For the waveform of this figure, Q is approximately 4.5.

transducer. In a real sense, the crystal acts in the same fashion as a bell when struck a sharp blow by a hammer, except that the losses due to radiation from the ultrasonic transducer are much larger than those found in a good-quality bell, so the transducer will not "ring" as long.

The rate of decay is proportional to the losses in the transducer, so it is natural to expect that the rate will be related to the Q of the crystal. Indeed, a definition of Q that is entirely consistent with Equation (5.4) but is of a different form can be given in terms of the energy lost per cycle of resonance as follows:

$$Q = \frac{\text{energy stored}}{\text{energy lost per cycle}} 2\pi \qquad (5.18)$$

If we let J represent the energy stored by the crystal, then Equation (5.18) can be rearranged in differential form as

$$\frac{dJ}{dt}\frac{1}{f_1} = -\frac{2\pi J}{Q} \qquad (5.19)$$

where f_1 = frequency of resonance. The solution to Equation (5.19) has an exponential decay as a function of time:

$$J = J'e^{-\gamma t} \qquad (5.20)$$

Substituting Equation (5.20) into Equation (5.19) and solving for γ yields the decay rate in terms of Q:

$$\gamma = \frac{2\pi f_1}{Q} = \frac{\omega_1}{Q} \qquad (5.21)$$

Since the power output of the transducer is proportional to the energy stored in its oscillations, and since the magnitude of the radiated pressure wave is proportional to the square root of the power in the wave, it is possible to write the time decay of the envelope of the pressure wave radiating from a transducer with a given Q as

$$\boxed{p = p_0 e^{-(\gamma/2)t} = p_0 e^{-(\omega_1/2Q)t}} \qquad (5.22)$$

where Equation (5.21) has been used to relate γ to the Q of the transducer. Thus, a high Q leads to a long ringing time whereas a low Q gives a shortened waveform. Figure 5.7 plots the pressure waveform for the example of a low-Q transducer.

As an approximate rule of thumb, it can be said that the number of cycles contained in the power waveform is roughly numerically equal to the Q of the transducer.* This can be shown by defining the point in time when the waveform is effectively ended to be that time t' when the power has diminished to $e^{-\pi} = 0.043$ of its original value and the pressure has

---

* As Problem 5.10 shows, "the Q of the transducer" is really not correct nomenclature since the value of Q is not a fixed characteristic of the transducer but is determined by the type of material against which the transducer is placed and will vary from application to application with the same transducer. However, in ultrasonic bioinstrumentation the transducer is invariably placed against tissue, so the resulting Q will be reasonable fixed.

5.2 ELECTRICAL EXCITATION OF PIEZOELECTRIC TRANSDUCERS

therefore diminished to $e^{-\pi/2} = 0.208$. This point is shown on Figure 5.7. From Equation (5.22),

$$\frac{\omega_1}{2Q}t' = \frac{\pi}{2}$$

so

$$t' = \frac{2\pi Q}{2\omega_1} = \frac{Q}{2f_1} \tag{5.23}$$

Since the period of one cycle of the power waveform is one-half the period of the pressure waveform (see Figure 3.1) and the pressure period is given by $1/f_1$, the period of the power waveform is $1/2f_1$, and

$$\boxed{t' \approx Q \text{ periods of power}}$$

As the rule of thumb states, there are approximately Q cycles of power (and, correspondingly, $Q/2$ cycles of pressure) contained in the pulse.

Axial Resolution

An important design question is now appropriate: Is it desirable to have a high-Q or a low-Q transducer for bioinstruments? The answer depends upon whether the instrument is operated cw (as some Doppler flowmeters are) or pulsed (as in echocardiography). If cw, for efficiency's sake it is best that the transducer has as high a Q as the transmission at the tissue interface will allow. The voltage exciting the transducer should then be a continuous sine wave centered at the resonant frequency of the crystal as determined by its thickness.

If operated pulsed, however, a low-Q transducer is desirable. This is because the *axial resolution* (*AR*) of the instrument is dependent upon the length of the pulsed waveform. Since the depth of the boundaries being investigated by a pulsed instrument is determined by measuring the round-trip transit time of the pulses reflected from the boundaries, the more accurate this time can be measured, the more accurate will be the determination of depth. It is clear that a shorter transmitted pulse will lead to a more precise measurement of the time of arrival of the echoes and, in turn, the depth of the reflecting borders. If we define the effective pulse time to be t' as previously and use the straightforward relationship that distance equals the product of time and velocity, we get

$$\text{Axial resolution} \approx \frac{t'c}{2} \tag{5.24}$$

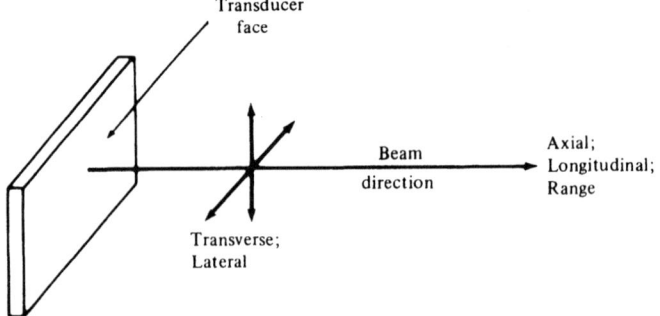

Figure 5.8 Conventions for directions related to beam propagation from a transducer.

where the factor of 2 enters because of the round-trip nature of the reflected wave (see Problem 5.6).

The nomenclature that is conventionally used to describe the directions related to a transducer and its propagating beam is summarized in Figure 5.8. Note that the terms "axial" and "longitudinal" are generally interchangeable, as are "transverse" and "lateral." Axial resolution pertains to spatial resolution in the direction of beam propagation, whereas transverse resolution is measured in the plane perpendicular to the beam's direction.

Seen in another way, axial resolution is a measurement of an instrument's ability to resolve two reflecting boundaries that are closely spaced in the axial (or longitudinal) direction of the instrument. Figure 5.9 shows the time sequence of pulses reflected from two closely spaced interfaces. It can be said that when the two boundaries are spaced apart in the longitudinal direction a distance equal to or greater than the axial resolution,* they can be resolved as separate reflectors. When they are closer, their echoes blend into one another.

Since the effective time length of a transducer's pulse is related to Q, Equation (5.24) for axial resolution can be rewritten using Equation (5.23):

$$\text{Axial resolution} \approx \frac{Qc}{4f_1}$$

Put in terms of wavelength,

$$\boxed{AR \approx \frac{Q\lambda}{4}} \qquad (5.25)$$

---

* Some authors prefer defining resolution as the *inverse* of minimal resolvable distance, with units of cycles per mm. In this text, we will use distance directly, since this definition seems more straightforward. In any case, the units tell the definition.

5.2 ELECTRICAL EXCITATION OF PIEZOELECTRIC TRANSDUCERS

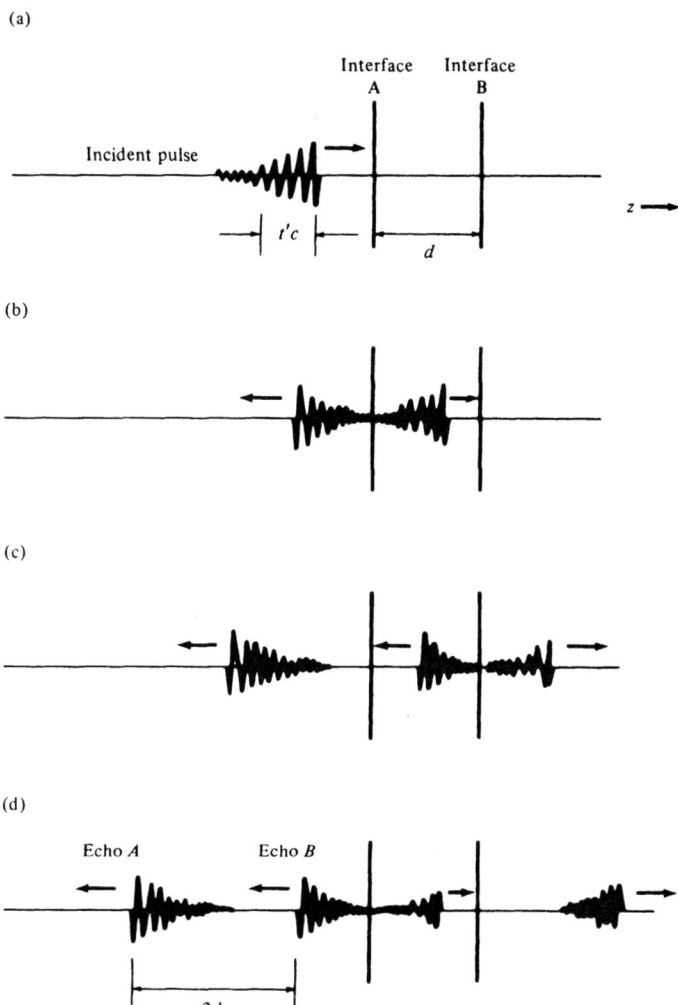

Figure 5.9 Four successive snapshots of the positions and lengths of echoes from two closely spaced interfaces. When d is reduced to the point where the echoes overlap but are just resolvable, then d = axial resolution.

which shows that improved resolution (a smaller value for Equation (5.25)) is a result of a lower-Q transducer. In fact, loss is sometimes purposely added to the back face of a transducer to lower its Q and improve its resolution. Although the total acoustic output power of the transducer (for a given electrical excitation) is reduced by this technique, the increased axial precision of imaging is often worth the cost. Figure 5.10 shows how

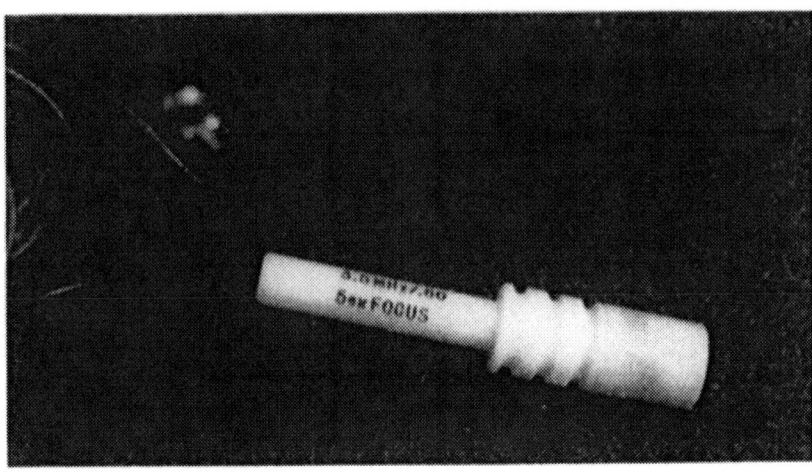

Figure 5.10 (a) A photo of a typical 3.5-MHz single element transducer. Its diameter is approximately 1.5 cm. (b) In order to reduce the Q and improve AR, some transducers have an absorber added on the rear face of the transducer.

this might be accomplished. Instead of air on the back side of the transducer, a material having an acoustic impedance much closer to the transducer's impedance is placed in close contact with the rear transducer face. This allows power to flow out the rear of the transducer in addition to that radiated into the tissue, thus lowering the transducer's Q. If the backing material is a good ultrasound absorber, this power is permanently lost. Absorber materials that have been successfully used include aluminum-filled epoxy and tungsten-filled epoxy. Problem 5.8 shows that an absorber-backed transducer will possess a lower Q and better AR than an air-backed one.

Equation (5.25) also reveals an important relationship between resolution and wavelength. The shorter the wavelength, the better will be the instrument's ability to resolve detail since a small value for axial resolution leads to improved measurement of spacings. For good resolution, an ultrasonic instrument should employ as high a frequency (as short a wavelength) as possible, limited only by the increased attenuation at the higher frequencies. For example, adult echocardiography is normally done at 2.25 MHz as a compromise between resolution and penetration. Pediatric echocardiography, however, will use frequencies as high as 5 MHz to improve resolution, since the path length into the heart is shorter in children and higher attenuation per centimeter is therefore allowed.

There are several practical factors that cause the actual axial resolution of a typical medical instrument to be worse than predicted by Equation (5.25). One is due to the frequency dependence of tissue absorption, called dispersive absorption, which will effectively lengthen the pulse as it travels through intervening tissue. As discussed in Chapter 4, most tissues show a linear increase in absorption with increasing frequency; high frequencies are attenuated much more than lower frequencies. A sharp pulse of transmitted acoustical energy (such as shown in Figure 5.7) actually contains a wide spectrum of frequency components, obtained by Fourier analysis of the time waveform of the pulse. The sharper the pulse, the higher the frequencies contained in its spectrum. (It can be said that the high-frequency components contribute to the "sharpness" of the pulse.) When this pulse travels through tissue, these higher frequencies are selectively lost at a faster rate than the low-frequency components are. The result is a stretching of the pulse time leading to worse axial resolution between neighboring reflectors.

Another factor is any electronic compression which may be purposely added in the receiver stages of the instrument to decrease its signal dynamic range before the display (covered in Chapter 6). Often, logarithmic compression is employed. Compression has the effect of minimizing the differences between large-amplitude signals and small-amplitude signals. When applied to the pulse waveform shown in Figure 5.7, it can be seen that the effect is to boost the tail of the pulse and therefore to effectively lengthen the pulse in time as seen on the display, again leading to a worsening of the axial resolution.

5.3 BEAM PATTERNS

We now turn our attention to the description of the shape of the radiating beam from the transducer. The behavior of this beam is important in determining the spatial sensitivity of the imaging instrument, both in the transmit and the receive modes.

The pressure wave that propagates from the face of an unfocused transducer generally maintains the approximate lateral dimensions of the transducer for a certain distance, but natural divergence begins to spread the transverse extent of the beam at larger distances so that the beam takes on a diverging nature. In the region near the transducer (the "near field"), the beam has many amplitude and phase irregularities due to interference between the contributing waves from all parts of the transducer's face, whereas in the region further from the transducer (the "far field"), the beam profile is much more uniform and well behaved. To quantitatively define the transition distance between these near-field and far-field regions, and to more precisely determine the amount of beam spreading in the far field, we next mathematically solve for the radiation pattern from an ultrasonic transducer.

The geometry of the problem is given in Figure 5.11; a circular coordinate system is initially assumed. The coordinates of the source points in the plane of the transducer face are denoted ρ and θ, and the coordinates pointing to the observation point where the pattern is sought are denoted r and ϕ. The distance from the source points to the observation point is given by r'. For circularly symmetric situations no generality is lost by letting the observation points lie on the x_1 axis. From geometry (see Problem 5.11),

$$r' = (r^2 + \rho^2 - 2r\rho \cos\theta \sin\phi)^{1/2} \qquad (5.26)$$

To analyze the observed radiation pattern, we rely upon Huygen's principle, which states that the radiation pattern from a general extended source can be constructed by considering the source as an appropriately weighted collection of point sources, each radiating outwardly propagating spherical waves. To get the complete radiation pattern, the contributions

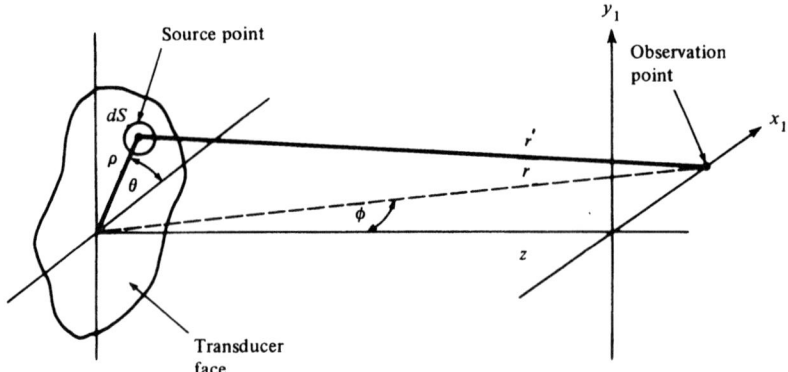

Figure 5.11 The general coordinates for solving for the radiation pattern from an ultrasound transducer.

5.3 BEAM PATTERNS

of all spherical waves from all point sources comprising the transducer are added (magnitude and phase) at the point of observation. This decomposition of the complex problem into a summation of simpler parts (i.e., spherical waves radiating from point sources) is allowed because the wave equation is a linear equation in the pressure variable, as shown in Problem 2.7.

Each point on the transducer face, then, is assumed to be the radiator of a spherical pressure wave, the form of which is

$$dp = \frac{kZu_0}{2\pi r'} \cos\left(\omega t - kr' + \frac{\pi}{2}\right) dS \tag{5.27}$$

where dp is the incremental pressure contribution at the observation point due to a spherical wave from a point source of incremental size dS, k is the propagation constant of the wave ($k = 2\pi/\lambda$), Z is the acoustical impedance of the intervening medium, and r' is the distance from source to observation point as given by Equation (5.26). Note that the pressure decreases as a function of $1/r'$ away from the point source; this is consistent with the $1/r^2$ dependence of power density expected from the conservation of energy principle applied to a diverging spherical wave. In obtaining Equation (5.27), the transducer face was assumed to be vibrating with a sinusoidal velocity of $u = u_0 \cos(\omega t)$ perpendicular to the ρ–θ plane.

The total pressure at the observation point is the integral of the incremental pressures:

$$p = \int_{source} dp \tag{5.28}$$

Assuming all portions of transducer face are oscillating with the same velocity and are in phase with each other, as would be the case for a rigid, pistonlike transducer, Equations (5.27) and (5.28) may be combined to give

$$p = \frac{kZu_0}{2\pi} \int_{source} \frac{\cos(\omega t - kr' + \pi/2)}{r'} \rho \, d\rho \, d\theta \tag{5.29}$$

where $\rho \, d\rho \, d\theta$ has been substituted for dS. For a general transducer shape and for an arbitrary observation point, this equation is quite difficult to evaluate and usually requires a computer solution. But it may be evaluated for some simple cases, as shown below.

5.3.1 Near-Field Pattern (On-Axis) of a Circular Transducer

Consider the transducer to be a circular disc of radius a. In the near-field region, r' is not large enough to allow a mathematical simplification of its form, so Equation (5.29) is still too complex for a general solution.

We therefore restrict our observation points to be on the z axis, such that $\sin \phi = 0$ and $r = z$. Equation (5.26) then reduces to

$$r' = \sqrt{\rho^2 + z^2}$$

Figure 5.12 shows the geometry for this special case. Substituting for r' in Equation (5.29) gives

$$p(z, t) = \frac{kZu_0}{2\pi} \int_0^a \frac{\cos(\omega t - k\sqrt{\rho^2 + z^2} + \pi/2)}{\sqrt{\rho^2 + z^2}} \rho \, d\rho \int_0^{2\pi} d\theta$$

Changing variables to $\beta = \sqrt{\rho^2 + z^2}$ and using straightforward integration leads to

$$\begin{aligned} p(z, t) &= -Zu_0[\sin(\omega t - k\sqrt{a^2 + z^2} + \pi/2) - \sin(\omega t - kz + \pi/2)] \\ &= Zu_0[\cos(\omega t - kz) - \cos(\omega t - k\sqrt{a^2 + z^2})] \end{aligned} \quad (5.30)$$

This result for the on-axis pressure amplitude has a very interesting interpretation. Note that the first term in the equation, $Zu_0 \cos(\omega t - kz)$, is just the familiar form for a pressure wave that appears to be coming from the center of the transducer, whereas the second term, $Zu_0 \cos(\omega t - k\sqrt{a^2 + z^2})$, which subtracts from the first, appears to be a wave coming from a point at the edge (radius = a) of the transducer. The combination of these two waves, with phases that change at different rates as z varies, provides the destructive and constructive interference pattern which produces the irregularities found in the near field.

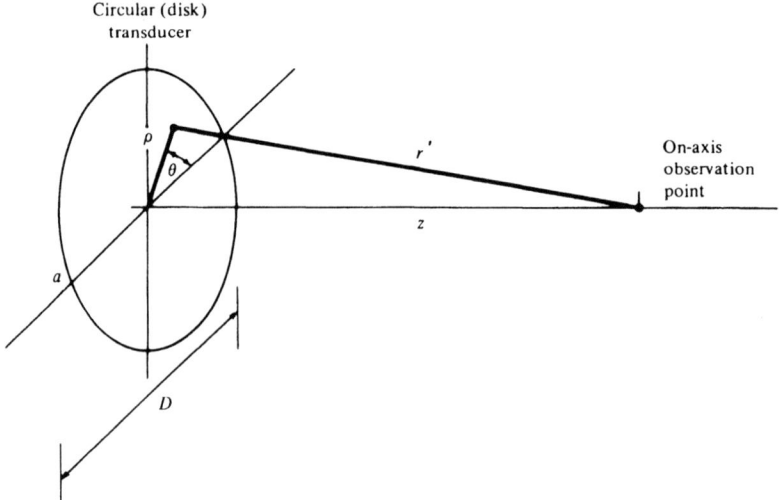

Figure 5.12 Geometry for calculating the near-field on-axis pressure field from a circular transducer of radius a and diameter D.

5.3 BEAM PATTERNS

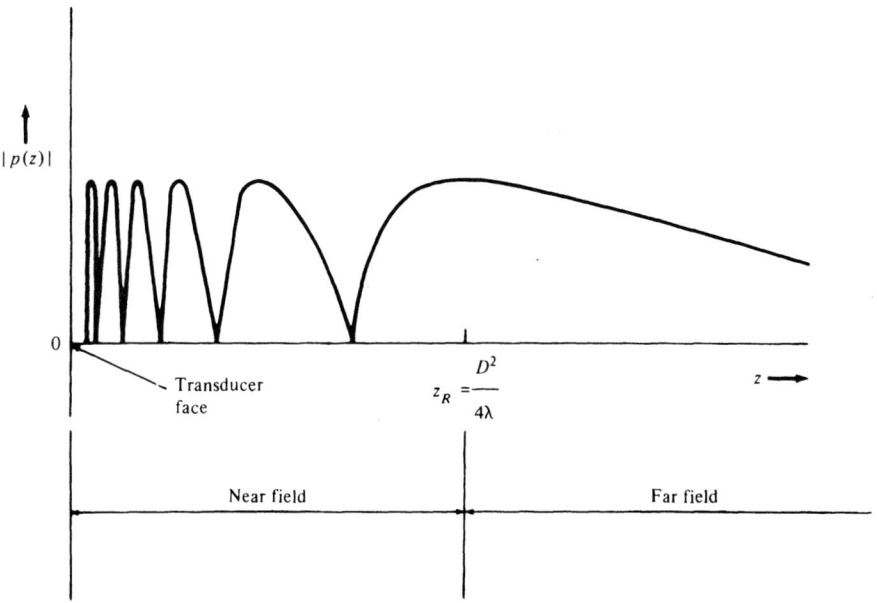

Figure 5.13 Variation of the magnitude of on-axis pressure field from a circular transducer of diameter D. This is a plot of the absolute magnitude of Equation (5.30) at one particular time, $t = 0$, and is the envelope of the oscillating pressure. The transition point from the near field to the far field is defined as the position of the furthest maximum. Beyond that point the field is more uniform.

A plot of the magnitude of Equation (5.30), shown in Figure 5.13, reveals the rapid variation of on-axis pressure in the near field of a circular transducer. Note that there is a multitude of points in the near field where the pressure actually goes to zero (complete destructive interference) and that the rapidity of the spatial oscillation of the pattern decreases as one moves further away from the face of the transducer. In fact, at large distances from the face of the transducer, the resultant pressure amplitude is no longer oscillatory but behaves as a slowly decreasing ($1/z$) field; this is the far field.

To mark the transition from near-field to far-field behavior, it is reasonable to choose the on-axis point where Equation (5.30) has its last maximum for increasing z. This is the value of z for which the phase difference between the first cosine term and the second cosine term in Equation (5.30) is just equal to π, so both terms are positive and the two terms add. Thus, if the transition point is denoted z_R, then

$$k\sqrt{a^2 + z_R^2} - kz_R = \pi \tag{5.31}$$

Since $z_R \gg a$ for a transducer many wavelengths in radius, the radical in Equation (5.31) may be approximated as

$$\sqrt{a^2 + z_R^2} \approx z_R + \frac{a^2}{2z_R}$$

Then Equation (5.31) becomes

$$kz_R + \frac{ka^2}{2z_R} - kz_R = \pi$$

Rearranging yields

$$\boxed{z_R = \frac{a^2}{\lambda} = \frac{D^2}{4\lambda}} \qquad (5.32)$$

For many unfocused transducers used in medical imaging, the body structures being imaged are not wholly in the far field of the radiation pattern where the fields are desirably uniform. For a 2-cm-diameter transducer at 2.25 MHz, the transition distance is $z_R = 15$ cm, rather deep.

If the transducer face is square or rectangular rather than round, the above equations do not strictly apply. For example, the pressure in the near field never goes exactly to zero anywhere as it does for a circular transducer. Nonetheless, the pressure magnitude has many peaks and valleys in the near field, and the qualitative description of the irregular near-field behavior making a transition to a more uniform far-field behavior is still valid. This is shown in Figure 5.14, where intensity maps (proportional to the square of pressure) are given at three progressively farther distances from a square transducer. The smoothing of the beam irregularities at greater distances is evident; however, even the most distant map shown in the figure is not yet in the far field of the transducer.

5.3.2 Far-Field Pattern of an Ultrasound Transducer

When the beam is observed at a large distance from the transducer, simplifications can be made in the general Equation (5.29) that allow the field to be calculated at any point (off-axis as well as on-axis) in the plane of observation. The approximations appear at two places in Equation (5.29). First, in the far field the magnitude of the r' term which appears alone in the denominator of the integrand will not differ appreciably from r over the range of the source integration, since r is much greater than the source dimensions. This r' is therefore set equal to r (a constant with respect to the integration variables) and is brought out of the integral.

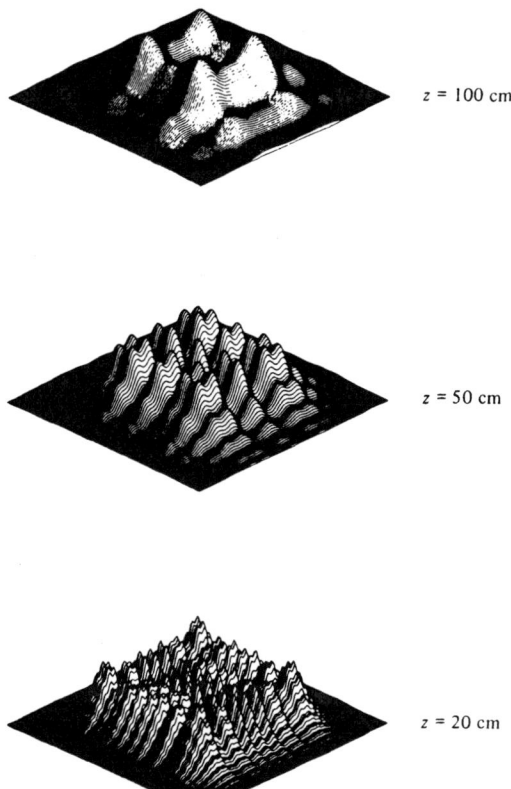

Figure 5.14 Radiation intensity maps at three planes progressively more distant from a square transducer face. Note the progression toward a more uniform distribution. The transducer is 5 cm × 5 cm operating at 3 MHz. Data from P. C. Pedersen and D. A. Christensen, *Acoustical Holography* 6 (1975), 711–739, Plenum Press.

Second, the r' term in the argument of the cosine also may be approximated, but since this r' is multiplied by k ($= 2\pi/\lambda$), and since λ is small at ultrasonic frequencies (making k large), this approximation cannot be as rough as letting $r' = r = $ constant; the phase term kr' may vary several radians over the source integration, causing several oscillations of the cosine term. To preserve this interference effect in the integration, only a partial simplification of r' from Equation (5.26) is made, known as the Fraunhofer approximation. If the observation distance is far enough away from the source that the source appears small compared to the observation length, then $r \gg \rho$ and Equation (5.26) may be approximated by

$$r' \approx r - \rho \cos\theta \sin\phi \qquad (5.33)$$

Putting Equation (5.33) into the phase term of Equation (5.29) and letting $r' = r$ in the denominator as discussed above gives

$$p(\phi, r, t) = \frac{kZu_0}{2\pi r} \int_{\text{source}} \cos\left[\omega t - kr + k\rho \cos\theta \sin\phi + \frac{\pi}{2}\right] \rho \, d\rho \, d\theta$$

$$= K \int_{\text{source}} \cos[\Psi(t) + k\rho \sin\phi \cos\theta] \rho \, d\rho \, d\theta \qquad (5.34)$$

where

$$K = \frac{kZu_0}{2\pi r}$$

and

$$\Psi(t) = \left(\omega t - kr + \frac{\pi}{2}\right)$$

K and $\Psi(t)$ are constants with respect to integration over the source coordinates ρ and θ. The result of this integration will depend upon the particular shape of the transducer. Two cases are considered next.

Circular Disk Transducer of Radius a

For a circularly symmetric source, the limits of integration become simply

$$p(\phi, r, t) = K \int_0^a \int_0^{2\pi} \cos[\Psi(t) + k\rho \sin\phi \cos\theta] d\theta \, \rho \, d\rho$$

Using the trigonometric identity $\cos(A + B) = \cos A \cos B - \sin A \sin B$ gives

$$p(\phi, r, t) = K \int_0^a \left[\cos\Psi(t) \int_0^{2\pi} \cos(k\rho \sin\phi \cos\theta) d\theta \right.$$
$$\left. - \sin\Psi(t) \int_0^{2\pi} \sin(k\rho \sin\phi \cos\theta) d\theta\right] \rho \, d\rho \qquad (5.35)$$

The last integral in this equation is zero since the sine term is an odd function of the cyclical argument ($\cos\theta$) as θ ranges from 0 to 2π. The other integral over θ is of the form which results in a Bessel function, as given in reference texts on Bessel functions:

$$\int_0^{2\pi} \cos(x \cos\theta) d\theta = 2\pi J_0(x)$$

5.3 BEAM PATTERNS

where J_0 is the Bessel function of the first kind with order zero. Therefore, Equation (5.35) becomes

$$p(\phi, r, t) = 2\pi K \cos \Psi(t) \int_0^a J_0(k\rho \sin \phi) \rho \, d\rho \qquad (5.36)$$

This integral, in turn, can be evaluated by the relation

$$\int x J_0(x) dx = x J_1(x)$$

where J_1 is the Bessel function of the first kind with order 1. Then, Equation (5.36) becomes (see Problem 5.13)

$$p(\phi, r, t) = \pi a^2 K \cos \Psi(t) \left[\frac{2 J_1(ka \sin \phi)}{ka \sin \phi} \right] \qquad (5.37)$$

The instantaneous radiated power density pattern may be found from $I = p^2/Z$, and reinserting the previous definitions for K and $\Psi(t)$ yields

$$I(\phi, r, t) = \frac{\pi^2 a^4 u_0^2 Z \sin^2(\omega t - kr)}{\lambda^2 r^2} \left[\frac{2 J_1(ka \sin \phi)}{ka \sin \phi} \right]^2 \qquad (5.38)$$

Some interesting observations about the far-field radiation pattern from a circular transducer may be obtained from Equation (5.38). First, note that the power density decreases as $1/r^2$ in this region, as would be expected when the measurements are made far enough away that the source appears as a small radiator of diverging waves. More importantly, the distribution with respect to angle behaves according to the term in the square brackets, the so-called directional factor:

$$H_c(\phi) = \left[\frac{2 J_1(ka \sin \phi)}{ka \sin \phi} \right] \qquad (5.39)$$

To obtain a feeling of the shape of this far-field pattern, it may be plotted on an observation screen a distance z away from the transducer. If the angles of divergence of the beam are not too great, the small angle approximation

$$\sin \phi \approx \frac{x_1}{z} \qquad (5.40)$$

may be used, where x_1 is the coordinate in the plane of observation; refer to Figure 5.11. Then, the directional factor Equation (5.39) may be written in terms of distance on the observation plane:

$$H_c(x_1) = \left[\frac{2 J_1(kax_1/z)}{kax_1/z} \right] \qquad (5.41)$$

The square of this term is plotted in Figure 5.15a and gives an indication of the extent of the power density pattern at a distance z from the transducer. Due to the denominator of $H_c(x_1)$ and the behavior of J_1, the power density drops off rapidly as x_1 increases from the center of the pattern. Also, there are repetitive zeros and side peaks as the Bessel function J_1 oscillates with the increasing arguments.

The great majority of power is contained in the central (main) lobe of the pattern between the first zeros on either side of this central peak. However, some power is found in the side lobes which neighbor the main lobe. The extent of the main lobe may be defined as occupying the area between the first zeros; these zeros occur at

$$J_1(\pm 3.83) = 0 \tag{5.42}$$

or

$$x_1 = \pm 3.83 \frac{z}{ka}$$

as shown in Figure 5.15a. Note that the width of the main lobe increases linearly with distance z in the far-field region.

Returning now to the angular dependence of the far-field radiation pattern, Equation (5.39) shows that the pattern may be considered to be a circularly symmetric function of the angle ϕ via the term $\sin \phi$; this equation is valid even for large ϕ. An angular plot of the logarithm of the square of Equation (5.39) in terms of decibels (to compress the range) is given in Figure 5.15b in polar coordinates; such a plot is sometimes referred to as the *antenna* pattern of the radiator. To obtain such a specific angular plot, a value of the transducer radius a must be given. For this figure, the transducer diameter is assumed to be 10 wavelengths wide, so $a = 5\lambda$ or $ka = 10\pi$.

The angular position of the first zero defines the amount of divergence (half-angle) ϕ_d of the main lobe as it propagates from the source; from Equation (5.42),

$$\sin \phi_d = \frac{3.83}{ka}$$

or

$$\boxed{\phi_d = \sin^{-1}\left(0.61 \frac{\lambda}{a}\right)} \tag{5.43}$$

It is convenient to use this angle as a measure of divergence of the beam from a circular transducer, although some authors consider it too conservative. The smaller angular width to the half-power points (-3 dB) rather than to the zeros is sometimes used; twice this angle is known as the Full

5.3 BEAM PATTERNS

(a)

(b)

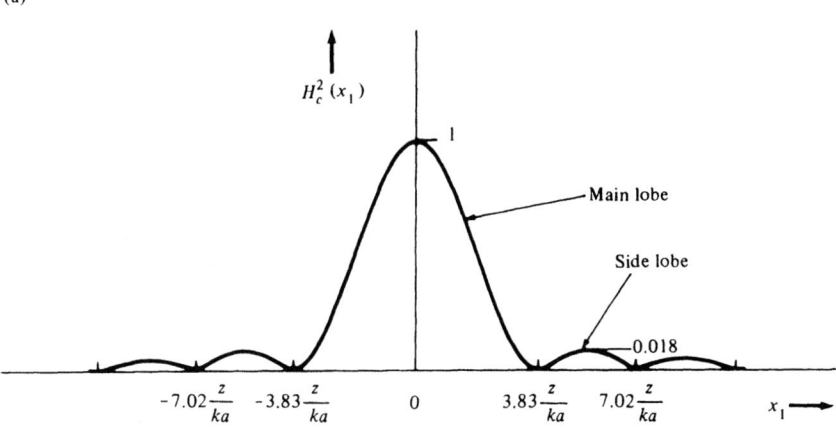
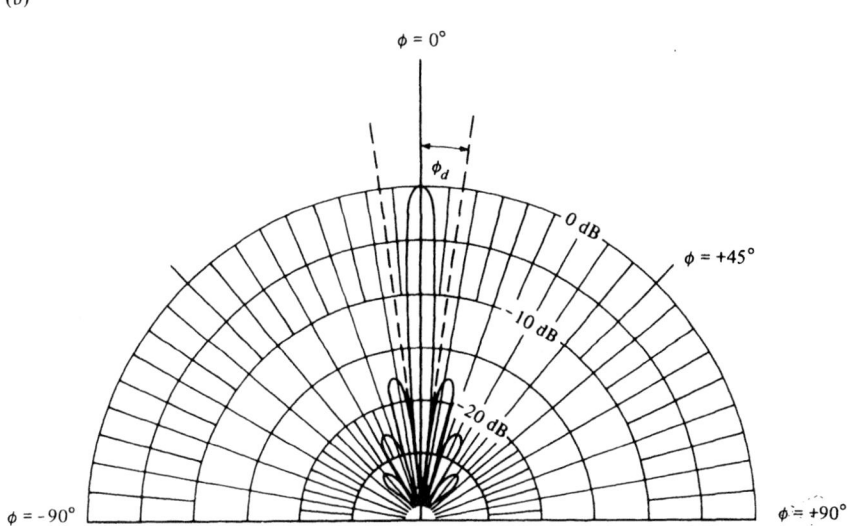

Figure 5.15 (a) The far-field power density pattern observed at a distance z from a circular transducer of radius a. (b) An angular plot of the same pattern in polar coordinates. The intensity is plotted in logarithmic (decibel) units. In this example, the transducer diameter equals 10 wavelengths.

Width to Half Maximum (FWHM) of power density. In this text, however, we shall use ϕ_d given by Equation (5.43) as the measure of divergence (half-angle) for reasons that will become clear when the concept of lateral resolution is discussed.

Note the inverse relationship in Equation (5.43) between ϕ_d and the transducer radius a. When a is a large number of wavelengths (as measured in the tissue), the far-field beam is highly directed; conversely, when a is small, the beam spreads considerably as it propagates from the transducer. In fact, when a is approximately one-half a wavelength (i.e., the diameter is one wavelength) or smaller, the half-angle of divergence is greater than 90° and the beam appears to be radiating hemispherically more or less isotropically from a point source.

A word of caution regarding the use of far-field patterns: Most single transducers used in medical imaging are many tissue wavelengths in diameter so the transition distance z_R is large enough for the reflecting objects to fall in the near-field region. Also, the transducers are often *focused* by an integral lens (covered in Section 5.5). In either case, the far-field divergence angle ϕ_d does not directly apply to the imaged region. However, multiple-element transducers, such as the linear arrays found in real-time scanners and discussed at the end of this chapter, are usually made of a series of small unfocused elements, and the radiation pattern from each of these small elements *is* determined by the far-field considerations of Equation (5.43). Also, for focused transducers the shapes of the beams in the focal plane will be shown later to be scaled-down versions of the far-field patterns found above.

Rectangular Transducer of Dimensions b × h

The analysis of the far-field radiation from a rectangularly shaped transducer with width b in the x_0 direction and height h in the y_0 direction proceeds from Equation (5.34) in a manner similar to that outlined above for a circular one; Figure 5.16a shows the orientation. Note that in the source plane, $\rho \cos \theta = x_0$. Initially restricting our observation to be along the x_1 axis ($\phi = \phi_x$), Equation (5.34) may be integrated over the rectangular source to give (see Problem 5.14)

$$p(\phi_x, r, t) = bhK \cos \Psi(t) \left[\frac{\sin[(kb \sin \phi_x)/2]}{(kb \sin \phi_x)/2} \right] \quad (5.44)$$

A similar expression holds for observations along the y_1 axis ($\phi = \phi_y$), and since the source is the shape of a rectangle whose boundaries may be expressed by equations that are mathematically separable in x_0 and y_0, the complete expression for far-field power density from a rectangular transducer is also separable in ϕ_x and ϕ_y:

$$I(\phi_x, \phi_y, r, t) = \frac{b^2 h^2 u_0^2 Z \sin^2(\omega t - kr)}{\lambda^2 r^2} \left[\frac{\sin[(kb \sin \phi_x)/2]}{(kb \sin \phi_x)/2} \frac{\sin[(kh \sin \phi_y)/2]}{(kh \sin \phi_y)/2} \right]^2$$

5.3 BEAM PATTERNS

(a)

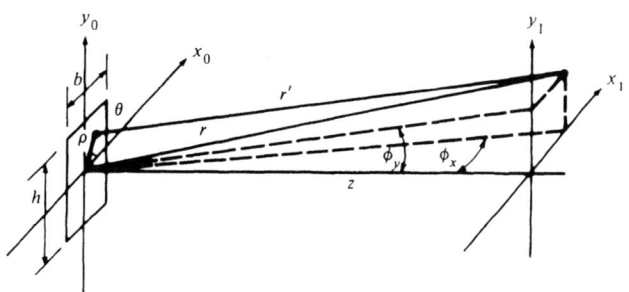

(b)

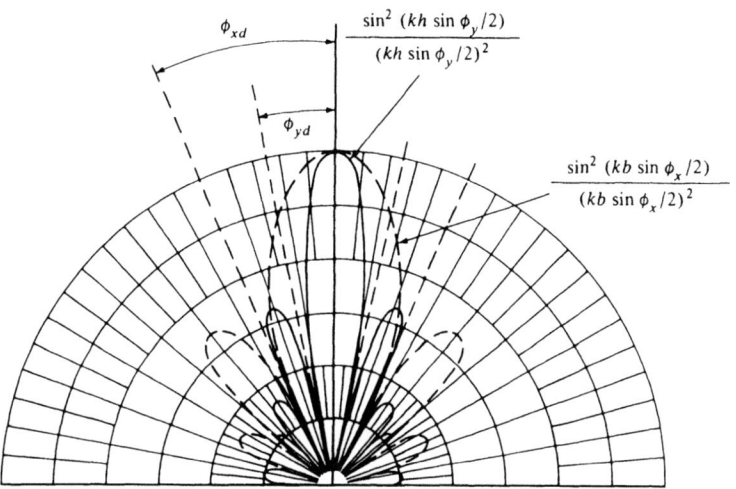

Figure 5.16 (a) The geometry for determining the far-field radiation from a rectangular transducer. (b) The far-field power density pattern as observed in the x_1 direction superimposed on the pattern in the y_1 direction.

The term in the square brackets is the directional factor:

$$H_r(\phi_x, \phi_y) = \left[\frac{\sin[(kb \sin \phi_x)/2]}{(kb \sin \phi_x)/2} \frac{\sin[(kh \sin \phi_y)/2]}{(kh \sin \phi_y)/2}\right] \quad (5.45)$$

The far-field beam pattern from a rectangular element has the same qualitative features as those described for a circular source, such as main lobe, side lobes, and so on, except that the directionality now has the form

$(\sin x)/x$. Note that the half-angle to the first zero marking the extent of the main lobe is now given in the x_1 direction by

$$\sin\left(\frac{kb \sin \phi_{xd}}{2}\right) = 0$$

or

$$kb \sin \phi_{xd} = 2\pi$$

or

$$\phi_{xd} = \sin^{-1}\left(\frac{2\pi}{kb}\right) = \sin^{-1}\left(\frac{\lambda}{b}\right) \tag{5.46}$$

A similar equation describes the divergence as measured in the y_1 direction:

$$\phi_{yd} = \sin^{-1}\left(\frac{\lambda}{h}\right) \tag{5.47}$$

Figure 5.16b shows that, as opposed to the pattern from a circular transducer, the pattern here is asymmetric. The inverse relationship between size and divergence angle still applies, however. For a rectangular element that is taller than it is wide (i.e., $h > b$), the far-field radiation pattern of the element will be wider than it is tall [i.e., $\phi_{xd} > \phi_{yd}$ from Equations (5.46) and (5.47)]. More will be said about rectangular radiation patterns when arrays of small elements are discussed at the end of this chapter.

5.4 WIDTH OF BEAM IN NEAR FIELD AND FAR FIELD

As the previous section described, the beam pattern in the near field has a very irregular interior, with many peaks and valleys, especially near the transducer face; Figure 5.14 showed this. The lateral extent of the near field is roughly confined to the size of the transducer, although it must be admitted that it is difficult to precisely define the edge of such an irregular field.

As the beam progresses into the far field, its topology becomes much more smooth, eventually evolving into a well-defined single main lobe with low-intensity side lobes as shown in Figure 5.15a. The edges of this beam now spread linearly with distance, and the width of the main lobe, as given by the half-angle ϕ_d to the first zero on each side, asymptotically diverges at a constant angle inversely proportional to the transducer diameter and therefore the near-field beam diameter.

5.4 WIDTH OF BEAM IN NEAR FIELD AND FAR FIELD

This progressive spreading of the beam width is diagrammed qualitatively in Figure 5.17. In the near field, also known as the Fresnel region, the beam is nearly collimated until it approaches the transition distance z_R. At this point the beam has started to spread a little. (Some authors define a "transition zone" here between the near field and the far field within which the beam changes its character from nearly collimated edges to diverging edges.)

As it travels into the far field, also known as the Fraunhofer region, the beam widens further and eventually approaches the constant angle of

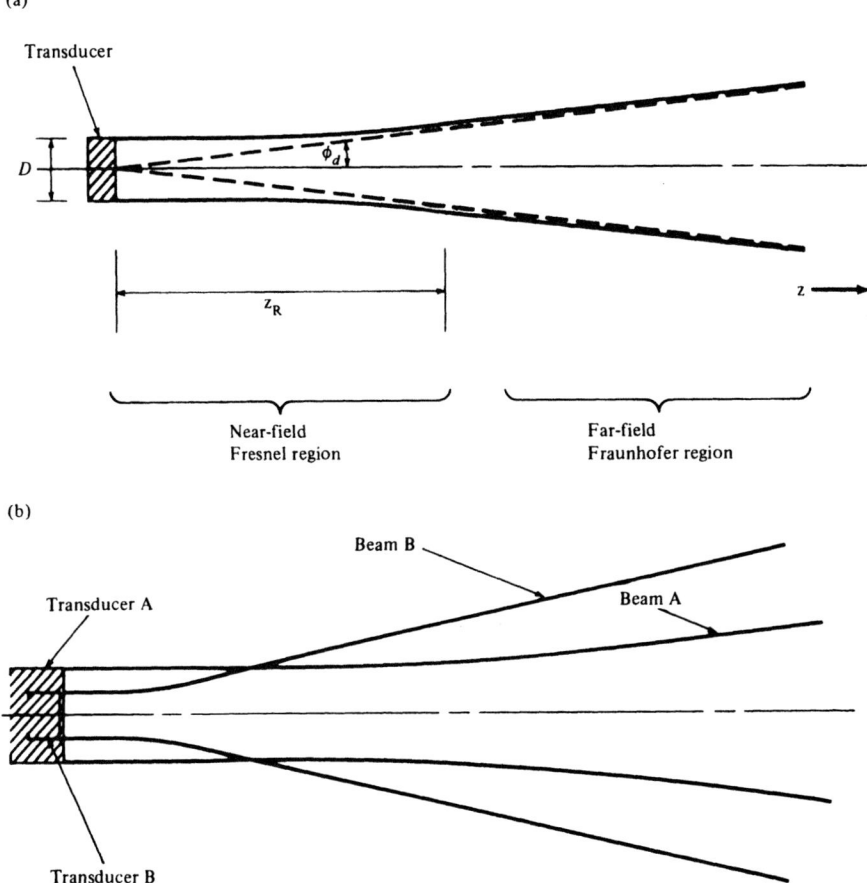

Figure 5.17 (a) The shape of a beam as it propagates away from an unfocused transducer. The beam stays approximately collimated in the near field, but diverges linearly in the far field approaching a half-angle ϕ_d. (b) The beam from a small transducer diverges more rapidly than the beam from a larger transducer.

divergence ϕ_d. It is interesting to note in Figure 5.17b that the beam from a small transducer (transducer B) starts small but eventually becomes larger since it possesses a shorter z_R and greater ϕ_d than transducer A.

Smoothing the Beam's Profile

The irregularities within the near field and the presence of side lobes in the far field are sometimes an inconvenience when attempting to predict the returns from reflectors in the beam's pattern. There is a technique for smoothing out these irregularities, although a price is paid. It is based upon the fact that, if the beam's amplitude profile as a function of radius was Gaussian-shaped at the transducer face (peaked at the center and decreasing to zero as $\exp(-\rho^2/a_1^2)$ toward the edges) rather than being the uniform amplitude across the transducer face assumed above, then the radiated beam's profile would be smoothly Gaussian-shaped *everywhere* in the near field as well as in the far field.

Therefore, if by some means the transducer excitation profile approximates a Gaussian form with decreasing activity away from the center, the beam would be expected to be more uniform in its transverse behavior. Various ways of achieving a shaded profile at the radiating surface include placing a radially varying absorber in front of the transducer, designing the transducer face to have a star shape with some unexcited areas near the edges, or by using a concentric ring transducer ("bull's-eye") and exciting the outer rings with progressively less drive voltage than the center rings. All these techniques, known as *apodization* because they reduce the "feet" (side lobes) in the radiated beam pattern, will produce an overall smoother beam profile. The disadvantages, however, are that less total power is radiated, the transducer is more complex, and, as Equation (5.43) shows, the beam diverges at a greater angle since the effective transducer diameter is smaller.

5.5 FOCUSING WITH LENSES, AND LATERAL RESOLUTION

The beam width from an unfocused transducer is generally too wide to give adequate definition of the fine lateral features of objects being imaged. Therefore, a lens or other focusing scheme such as a spherical reflector is usually employed to converge the radiating beam into a spot at the focal plane of the lens. However, the size of the focused beam cannot be infinitely small, since the natural divergence of a propagating wave as described in

5.5 FOCUSING WITH LENSES, AND LATERAL RESOLUTION

the previous section will attempt to spread even a converging beam, reducing the focusing effect of the lens. The further away the focal plane is from the lens, the larger the focused spot will be.

The equations of Section 5.3 can be used to evaluate the size of the focused spot once they are modified to include the effects of the lens. As in optics, an acoustical lens is fabricated from a disk of material by forming a curved refracting surface on one or both of its faces; Figure 5.18 shows the cross-section of a plano-concave focusing lens. As opposed to optics,

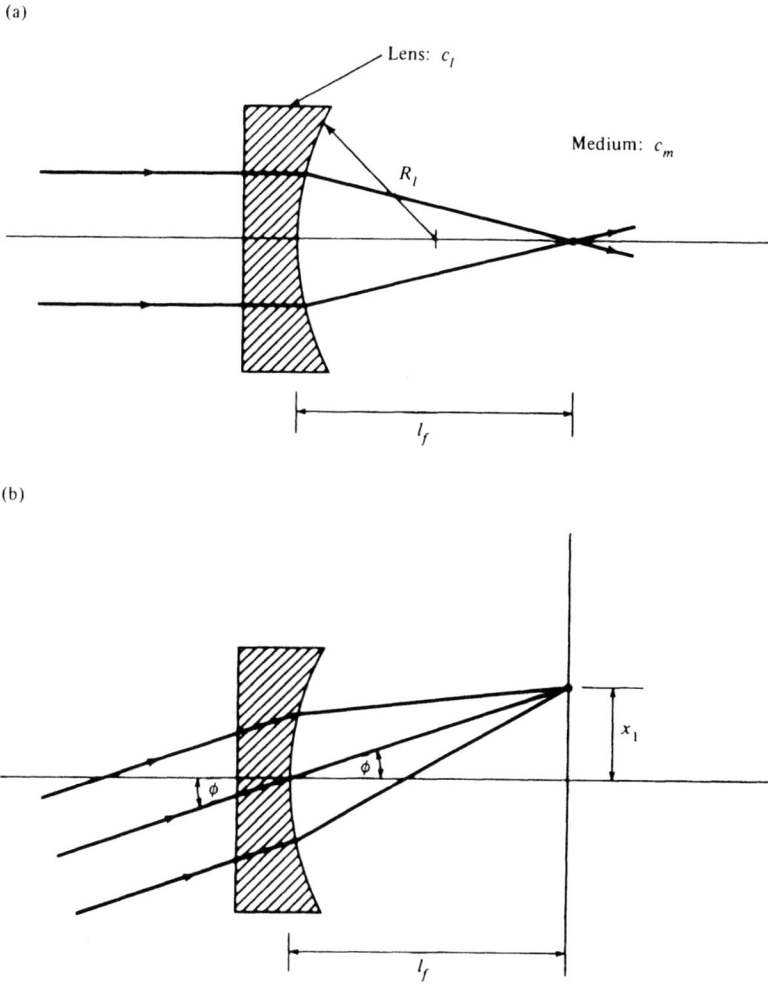

Figure 5.18 (a) A focusing lens made of material (such as polystyrene) with phase velocity greater than in the surrounding medium. (b) A lens has the property of transforming angles into position on the focal plane.

the lens material generally possesses an acoustic phase velocity which is *greater* than that of the material surrounding it (water or tissue). Thus, a converging (positive) lens will have a concave face. Problem 5.18 uses Snell's law for ray tracing to show that a plano-concave lens having a surface with radius of curvature R_l will produce focusing at a focal length equal to

$$l_f = \frac{R_l}{1 - \frac{c_m}{c_l}} \tag{5.48}$$

where c_l is the phase velocity of the lens material and c_m is the phase velocity of the medium into which the wave is focused.

Lenses have the property of transforming angles into position. That is, all rays entering the lens at a common angle ϕ will get directed to a radius x_1 on the focal plane as shown in Figure 5.18b. Under the small-angle approximation, geometry gives the transformation relationship as

$$\sin \phi \approx \frac{x_1}{l_f} \tag{5.49}$$

Therefore, a lens of focal length l_f placed in front of the beam from a circular transducer whose radiation pattern is given by Equation (5.39) will transform the far-field angular distribution into a spatial distribution on the focal plane via the transformation of Equation (5.49). Making this substitution into Equation (5.39) yields the spatial distribution of the pressure at a focused spot from a circular transducer of radius a:

$$H_c(x_1) = \left[\frac{2 J_1(kax_1/l_f)}{kax_1/l_f} \right] \tag{5.50}$$

and the *focused* pattern looks exactly like the far-field pattern of Figure 5.15a, except that it is scaled down by an amount l_f/z. Figure 5.19 shows how the focused spot would appear face-on.

Size of Focused Spot

The focused spot has a dense central portion (corresponding to the main lobe) surrounded by minor rings (the side lobes). The diameter of the central portion is defined as previously: the distance between the first zeros bounding the main lobe. From Equation (5.50) the radius of the first zero is found at

$$x_1 = \frac{3.83 l_f}{ka} = \frac{0.61 l_f \lambda}{a}$$

5.5 FOCUSING WITH LENSES, AND LATERAL RESOLUTION

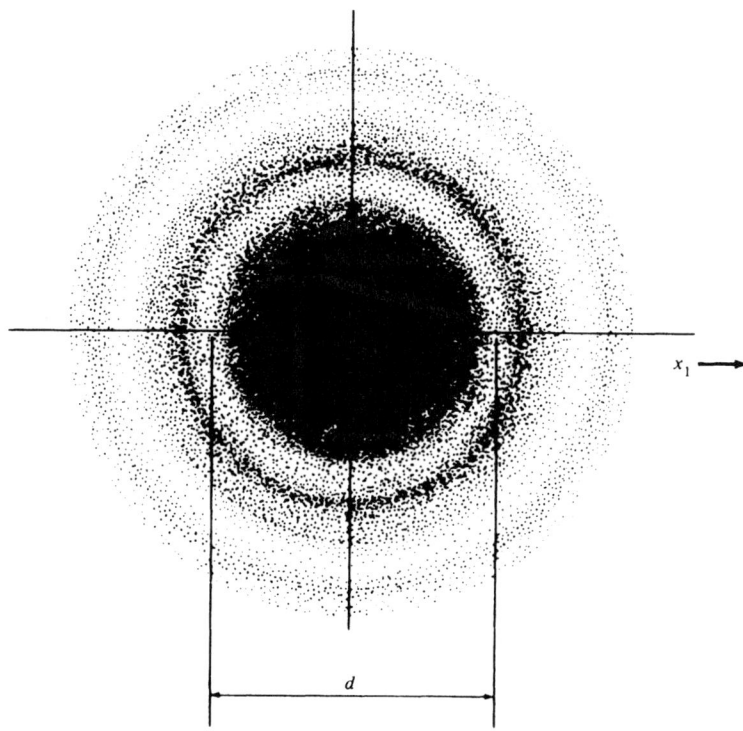

Figure 5.19 The greatly magnified pattern of a focused spot from a lens of focal length l_f. The entering beam diameter is D.

or, put in terms of original beam (transducer) diameter $D = 2a$, the diameter between first zeros is the focused spot diameter $d = 2x_1$:

$$d = 2.44 \left(\frac{l_f}{D}\right)\lambda \qquad (5.51)$$

Figure 5.20 defines the quantities entering Equation (5.51).

For the case of a rectangular transducer of width b, an analogous development can be undertaken to find the width w of the focused spot in the direction parallel to b. Using Equations (5.44) and (5.49), the result for a rectangular transducer is

$$w = 2\left(\frac{l_f}{b}\right)\lambda \qquad (5.52)$$

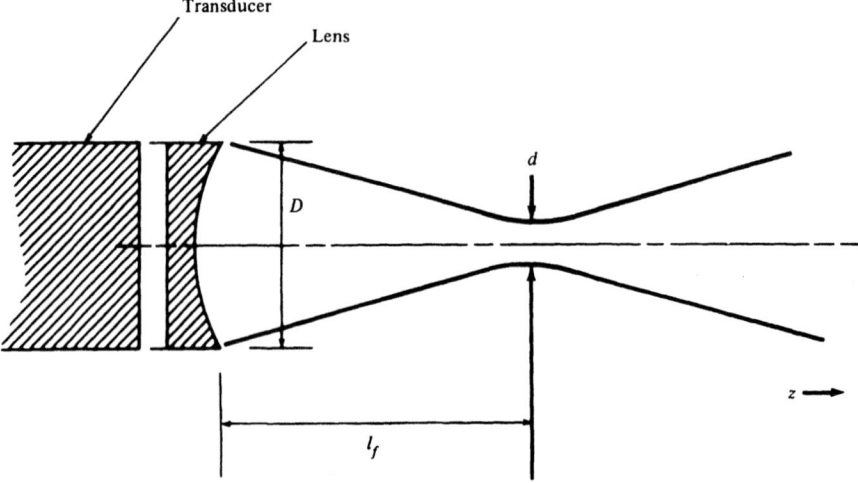

Figure 5.20 A lens will focus the beam to a small spot, but the size d of the focused spot depends upon l_f, D, and λ according to Equation (5.51).

The effects of divergence are manifested in these relationships. In Equation (5.51) the larger the diameter D of the transducer, the smaller is the tendency of divergence to expand the beam, and the smaller the spot of focus. Also, the further away the position of focus l_f, the greater is the effect of divergence, leading to a larger d.

How small can the beam be focused practically? The ratio in the parentheses in Equation (5.51) is known as the "f-number" of the lens; and due to practical limitations such as spherical aberration, it is difficult to fabricate a quality lens whose f-number is much smaller than unity. Therefore, as a rule of thumb, it can be said that the smallest possible focused spot is on the order of the wavelength of radiation used.*

Not only is it impossible to focus to an infinitely small spot, it may be impossible to get any narrowing at all in the beam diameter if the attempted focal distance is too far away. Equation (5.51) shows that d will be greater than D if

$$l_f > \frac{D^2}{2.44\lambda} \approx z_R \qquad (5.53)$$

or, in other words, no focusing occurs if the focal length of the lens is greater than about the transition distance. Thus, it may be said that focusing is only possible at distances within the near field, not in the far field of a transducer.

* This limitation appears in optics and general quantum-mechanical wave analyses as well as in acoustics.

Lateral Resolution

The spot size of the focused beam determines the transverse spatial resolution of a medical ultrasound imager, as indicated in Figure 5.21. Here, the focused beam is swept uniformly past a pair of point reflectors. The waveform of the envelope of the received echoes depends upon the lateral spacing of the points. When far apart, the echoes from each point are distinct, and it is clear that there are two separate points. As they move closer, however, approaching the spacing d, the separate echoes start to blend together, and at some stage the points are so close that their echoes cannot be separately resolved: that is, they appear as one reflecting object.

The spacing in the transverse plane at which the points are just separately resolvable is known as the *lateral resolution* (LR); and from Figure 5.21 a reasonable measure of LR is the diameter of the focused spot d. That is,

$$LR \approx d$$

and the motivation for focusing the beam to reduce the focused spot size d in an imaging system is obvious. Using Eq. (5.51) for a circular transducer,

$$\boxed{LR \approx 2.44\left(\frac{l_f}{D}\right)\lambda} \tag{5.54}$$

This relationship, along with the previous expression for axial resolution, is restated in Table 5.2.

We can now make an important observation: The resolution in all directions (axial and lateral) is closely related to wavelength, and as a practical matter cannot be made smaller than the wavelength used. Therefore, high resolution machines will employ as high a frequency as possible, until increasing attenuation takes the signal to the lower limit of the signal-to-noise ratio. Echo instruments for imaging tiny objects in the eye, for instance, may go as high as 15 MHz since the absorbing path length is so short there.

Depth of Focus

There is one disadvantage to tight focusing of the beam. Although it improves lateral resolution for reflecting objects located in the plane of focus, points in planes either nearer or further away than the focal length are compromised because the beam is somewhat larger than d on either side of the focal plane. The problem gets worse with decrease in the focused size, as shown in Figure 5.22.

The axial distance over which the beam maintains its approximate focused size is termed the depth of focus. To obtain an estimate of this

(a)

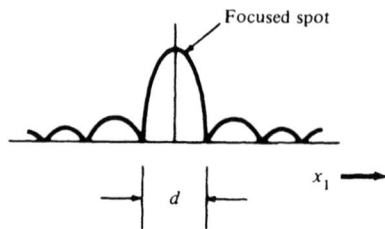

(b)

Reflector spacing: Received signal:

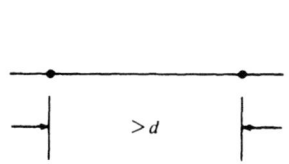 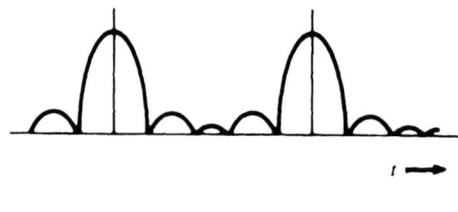

(c)

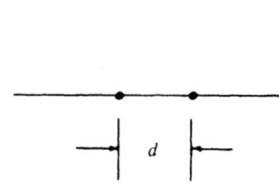 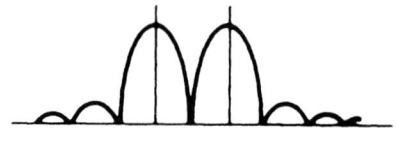

(d)

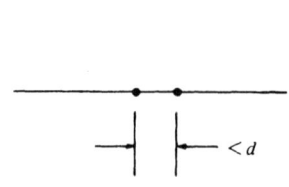 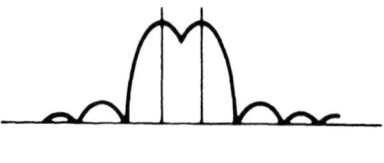

Figure 5.21 (a) The lateral spatial resolution is determined by the size d of the focused spot. (b)–(d) The signal received from a pair of point reflectors as their lateral spacing is progressively narrowed.

5.5 FOCUSING WITH LENSES, AND LATERAL RESOLUTION

TABLE 5.2 THEORETICAL EXPRESSIONS FOR SPATIAL RESOLUTION

$$AR \approx \frac{Q}{4}\lambda$$

$$LR \approx 2.44\left(\frac{l}{D}\right)\lambda$$

distance, note that the beam shapes on both sides of the focal point are mirror images of one another, reflected about the focal plane. The beam behavior on one side of this plane, from the focus outward, has the same general characteristics that we earlier examined in Figure 5.17a for a beam propagating from an initially planar wavefront of a given diameter. Therefore, it stays approximately collimated within the transition distance z_R. Applied to the situation here, the transition distance of Equation (5.32) becomes

$$z_{R,d} = \frac{d^2}{4\lambda}$$

The depth of focus may be estimated to be twice this distance due to symmetry about the focal plane:

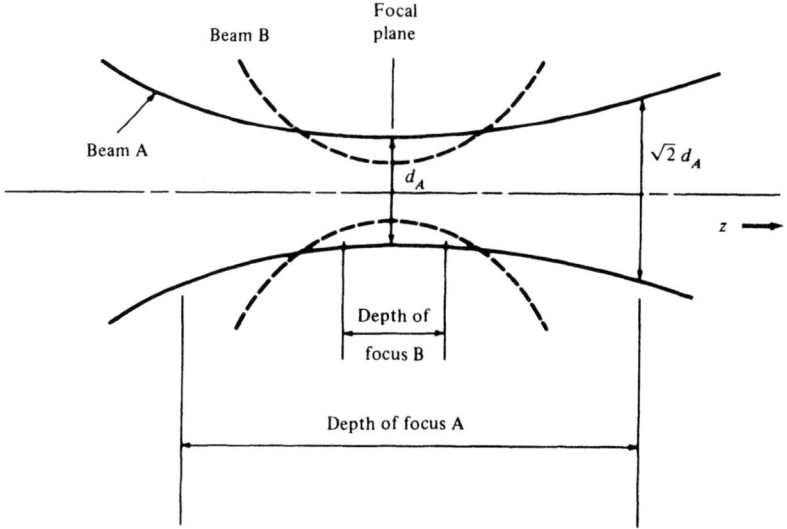

Figure 5.22 The depth of focus describes the longitudinal distance over which the beam maintains its approximate focused size. It gets shorter for tightly focused beams.

$$\boxed{\text{Depth of focus} \approx \frac{d^2}{2\lambda} \approx 3\left(\frac{l_f}{D}\right)^2 \lambda} \qquad (5.55)$$

where Equation (5.51) was used to get the last result.

The tradeoff between focused size and depth of focus sometimes dictates a compromise in lens design. For example, in fixed-focus systems, the lens may be purposely given a nonspherical surface to cause the focused spot size to be larger than the theoretical limit, thereby increasing the depth of focus.

Example of Resolution Values

A numerical example of the spatial resolution limits for a typical medical ultrasound system is enlightening. The transducer pictured earlier in Figure 5.10 is designed for shallow cardiac imaging. It operates at 3.5 MHz with a Q of approximately 7. Its beam diameter is 1.5 cm and the focal length of the lens is 5 cm. Therefore, $\lambda = 0.043$ cm in tissue, and the theoretical resolution limits are

$$AR \approx \left(\frac{7}{4}\right)0.043 = 0.075 \quad \text{cm}$$

$$LR \approx 2.44\left(\frac{5}{1.5}\right)0.043 = 0.35 \quad \text{cm}$$

and depth of focus is approximately equal to 1.4 cm.

However, practical factors will worsen the lateral resolution. These factors include lens aberrations, objects being outside the depth of focus, side-lobe (or grating-lobe) off-axis sensitivity, spatial interference noise (speckle) due to the coherent nature of ultrasound, and signal compression in the receiver electronics. In addition, as mentioned in Section 5.2.2, practical factors such as dispersive absorption in tissue and signal compression in the receiver will degrade the axial resolution. Therefore, the actual resolutions are perhaps two to three times their theoretical values above. Note that, as is usually the case, the axial resolution is much better than the lateral resolution.

5.6 LINEAR ARRAYS

In real-time B-scanners, described in the next chapter, the transducer is sometimes composed of a linear array of closely spaced elements (usually

5.6 LINEAR ARRAYS

rectangular) as shown in Figure 5.23. Although each element may be small in terms of number of wavelengths, the overall width L can be appreciable (5 to 10 cm). The question is: What width is used in Equation (5.46) to calculate the angular divergence of the beam radiated from this unfocused transducer, the single-element width b or the overall-array width L? The answer depends upon whether the elements are excited one at a time or whether they are all radiating together.

If excited one at a time (as is done in the sequentially pulsed linear array machine), the pattern is, not surprisingly, just that of a single element; this pattern was covered in Figure 5.16b and is relatively broad in the horizontal direction, or azimuth, due to the smallness of the elements.

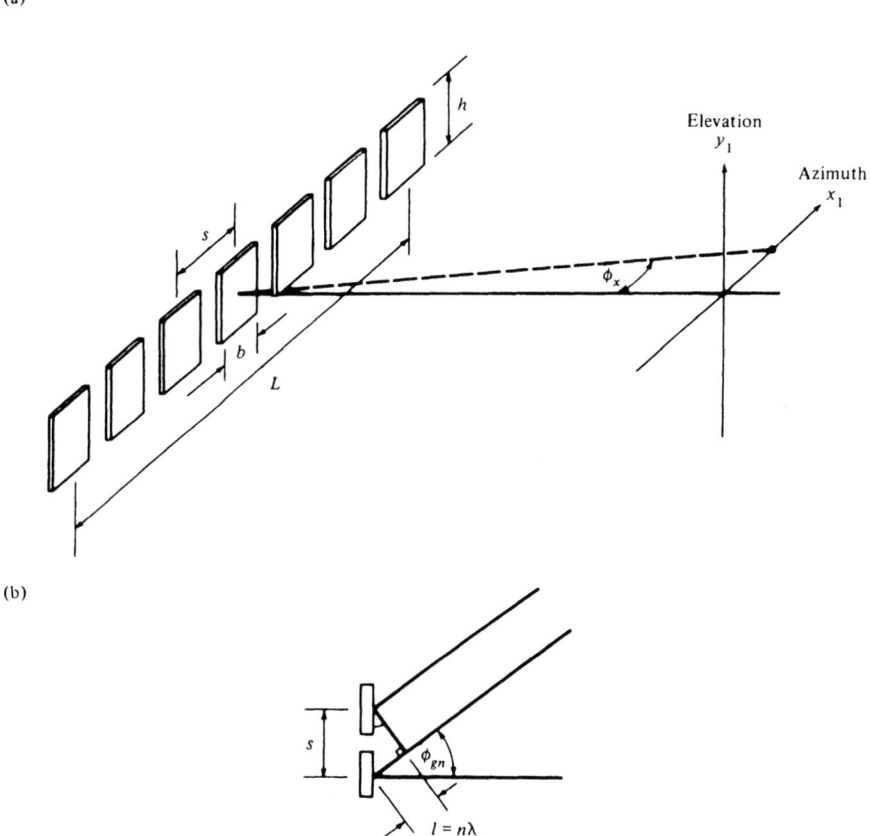

Figure 5.23 (a) The geometry for a linear array of rectangular elements. (b) A top view of two neighboring elements showing that at selected angles ϕ_{gn}, the path length difference l is equal to an integral number of wavelengths.

If the elements are excited simultaneously and coherently (as in the phased-array imager), the effective transducer width is L and the far-field divergence, given by Equation (5.46), will be much narrower. Correspondingly, if this coherent array is focused, the beam will converge to a smaller spot size with improved lateral resolution. (Note, however, that the beam divergence in the vertical direction, called the elevation, will be the same whether the elements are excited independently or coherently; in both cases the effective array height is h).

Grating Lobes

There is a complexity in the radiation pattern which accompanies the segmentation of the transducer into an array of elements. It is the appearance of reduced-amplitude images of the main beam (complete with side lobes) known as *grating lobes*, centered around one or more discrete angles in the ϕ_x plane. The angles of the grating lobes, denoted ϕ_{gn}, are found to be those angles for which rays from two neighboring elements are in phase with each other by a multiple of 2π; constructive interference therefore takes place at these angles, and some power is radiated in those directions. An alternate way of stating the condition for constructive interference is that the path length difference l between rays from the neighboring elements is equal to an integer number of wavelengths. Figure 5.23b shows that this occurs when

$$\sin \phi_{gn} = \frac{l}{s} = \frac{n\lambda}{s}$$

or

$$\boxed{\phi_{gn} = \sin^{-1}\left(\frac{n\lambda}{s}\right) \qquad n = \pm 1, \pm 2, \ldots} \qquad (5.56)$$

There will be as many grating lobe orders in the pattern as the number of solutions of Equation (5.56) that fall within $\pm 90°$. Notice that, as the spacing s increases in size with respect to a wavelength, the grating lobes get closer together in angle and increase in number.

Figure 5.24 shows an example of a 16-element array with a total length of $L = 27\lambda$. Therefore, $s = (27/15)\lambda = 1.8\lambda$ and

$$\phi_{g1} = \pm 33.7°$$

Only the first-order grating lobes are present in this pattern since the second-order ($n = 2$) and higher-order lobes are not valid solutions (within $\pm 90°$) of Equation (5.56) for $s = 1.8\lambda$. The shape of the main beam and the displaced grating lobes is determined by applying Equation (5.45) with the

5.6 LINEAR ARRAYS

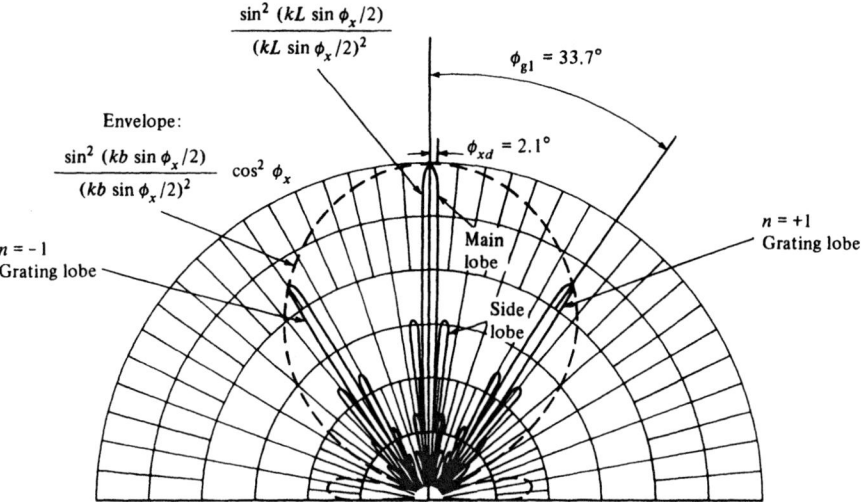

Figure 5.24 The power density beam pattern in the x_1 direction from a 16-element linear array whose elements are excited coherently. In addition to the main beam, copies of the main beam called grating lobes appear off to each side. For this example, $s = 1.8\lambda$ and $L = 27\lambda$.

effective length $b \to L$. Therefore, the width of the main lobe in the main beam (straight ahead) and also in the grating lobes is given by Equation (5.46):

$$\phi_{xd} = \sin^{-1}\left(\frac{\lambda}{L}\right) = \sin^{-1}\left(\frac{1}{27}\right) = 2.1°$$

The envelope which determines the amplitude of the grating lobes compared to the main beam is given by the directional factor H_r of one of the *individual* (assumed identical) elements, multiplied by a factor $\cos \phi_x$ (which is due to the lack of a reinforcing rigid baffle surrounding the elements and which has a major effect only near $\pm 90°$). Therefore, using Equation (5.45) for H_r,

$$\text{Envelope (amplitude)} = \frac{\sin[(kb \sin \phi_x)/2]}{(kb \sin \phi_x)/2} \cos \phi_x \qquad (5.57)$$

since each element has a width of b (refer to Figure 5.23a). This envelope will possess zeros just like the main-beam directional factor, but they will be at much larger angles since $b \ll L$. The positions of the zeros of the envelope (in addition to $\pm 90°$ from the $\cos \phi_x$ term) may be obtained from Equation (5.46):

$$\phi_{xe} = \sin^{-1}\left(\frac{m\lambda}{b}\right) \qquad m = \pm 1, \pm 2, \ldots \qquad (5.58)$$

Note that the closer to unity the ratio b/s is ("fill factor"), the closer the grating lobe angle from Equation (5.56) will be to a zero angle of the envelope from Equation (5.58), thus reducing the peak amplitude of the grating lobe. For the example of Figure 5.24, $b = 0.67s = 1.21\lambda$, and the envelope has the shape shown with zeros at $\pm 56°$.

An interesting envelope occurs for the special case of $s = 2b$ corresponding to a "square wave" array whose element's active width is just one-half the spacing between element centers. For this case, the angle of the second grating lobe

$$\phi_{g2} = \sin^{-1}\left(\frac{2\lambda}{s}\right) = \sin^{-1}\left(\frac{\lambda}{b}\right)$$

falls at the first zero in the envelope

$$\phi_{xe} = \sin^{-1}\left(\frac{\lambda}{b}\right)$$

and the second grating lobe essentially vanishes. In fact, all even-order grating lobes disappear, leaving only the main beam and odd-order grating lobes.

Grating Lobe Reduction

Reduction in the number and amplitude of the grating lobes is desirable since grating lobes represent potential sources of ambiguity in determining the direction of the echoes returned to the transducer. The principle of reciprocity applies under most circumstances to acoustic wave propagation, so the transducer's receiver sensitivity pattern will usually have the same shape as the transmitter radiation pattern. Therefore, grating lobes (and, to a lesser degree, side lobes) give off-axis sensitivity to a transducer used both as source and receiver. Reflecting points at the grating lobe angles are irradiated, and the receiver is sensitive to echoes coming from these angles in addition to straight ahead. Figure 5.25 shows how a single point scatterer will show up at three separate angular positions of a swept array with two grating lobes in addition to the main beam.

The angles of the grating lobes are governed by the spacing s between elements of the array, and their amplitude is determined by the envelope shape set by the individual element length b. In addition, the overall width L determines the angular width of each lobe, and the number of elements is given by $(L/s) + 1$. These array parameters can be manipulated by the

5.6 LINEAR ARRAYS

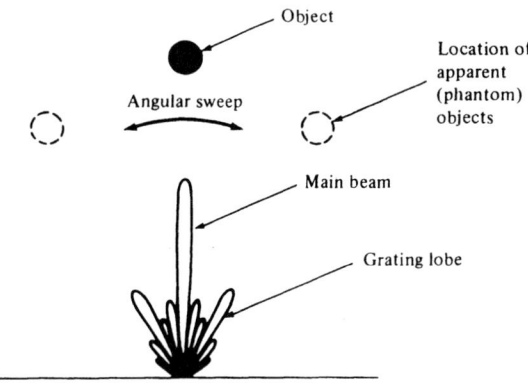

Figure 5.25 When the transducer is rotated, the grating lobes produce multiple responses from a single object, confusing the interpretation of object position.

designer to optimize one feature or another, depending upon the desired application, but they all interact. For example, to move the grating lobe angle as far away from 0° as possible, the spacing s between elements is made small. However, for a fixed number of elements, this reduces the array width L, which in turn *increases* the angular width of the main beam, worsening lateral resolution. Problem 5.22 gives other examples of the tradeoffs encountered in array design.

There is a temporal way to partially reduce the magnitude of the grating lobes in the transmitted pattern; it is based upon using very short transmitter pulses. As explained above, grating lobes are due to constructive interference occurring at selected angles between waves from neighboring elements. If the waves are really pulses of short duration (little more than one cycle), the pulse from one element propagating at the grating lobe angle will have decayed considerably by the time it is joined by the pulse from its neighbor, amounting to less than total constructive interference; this is diagrammed in Figure 5.26. When pulses from all the elements are considered, the skew in timing may significantly reduce the grating lobe response. The pattern in the forward direction remains essentially unaltered, however, since all pulses coincide in this direction, providing total constructive interference.

Thus, the ratio of grating lobe response to main lobe response decreases with decreasing pulse length (and therefore with decreasing transducer Q). For example, for a 16-element array with an interelement spacing of $b = 2.4\lambda$, the ratio of first grating lobe amplitude to main lobe amplitude is 0.2 when $Q = 9.4$ but only 0.08 when $Q = 3.1$. Unfortunately, this reduction in peak grating lobe response is accompanied by an increase in the angular width of the grating lobe.

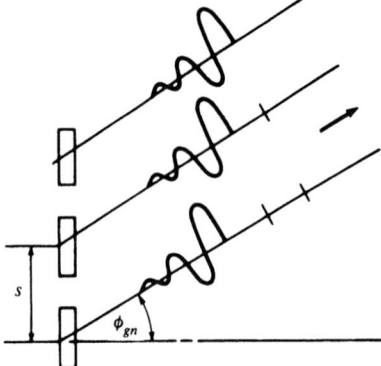

Figure 5.26 If short pulses are emitted from the elements of a linear array, their origins are skewed when summed at an angle. At the grating lobe angle ϕ_{gn}, some constructive interference takes place, but it is less than for continuous waves due to the decaying envelopes summed from neighboring pulses.

There is yet another way to attempt to reduce grating lobe response (for fixed L and for a given number of elements). The spacing between elements can be made nonuniform, defeating some of the constructive interference effects at off-axis propagation angles. This randomization of element spacing, however, is of minor benefit (especially for short pulses) and also increases the width of the grating lobe, so it is probably not worth the effort.

PROBLEMS

5.1. What is the thickness of a barium titanate transducer whose fundamental frequency of resonance is 5 MHz? How thick would it be if its third harmonic were 5 MHz?

5.2. A cw voltage with a peak magnitude of 10 V is impressed across a barium titanate transducer at its fundamental frequency of 1 MHz. The area of the transducer face is 1 cm². How much power is radiated into a layer of muscle in contact with the transducer?

5.3. Using the configuration shown in Figure 5.4, derive Equation (5.6) for the velocity of the transducer faces when excited at resonance by an electric field E_i. (*Hint:* Let the four traveling waves be displacement waves of the form

$$\xi_1 = A \cos(\omega t + kz)$$
$$\xi_2 = B \cos(\omega t - kz)$$
$$\xi_3 = C \cos(\omega t + k_1 z)$$
$$\xi_4 = D \cos(\omega t - k_2 z)$$

where ξ is the displacement of the material's particles from equilibrium and k, k_1, and k_2 are the propagation constants in the transducer, region 1, and region 2, respectively. Particle velocity u is given by $\partial \xi / \partial t$. Match particle velocity u and pressure p at each interface following the sign conventions used in Section 3.4.2. The pressure just inside the transducer face is given by Equation (5.5) for a piezoelectric material. The net displacement inside the transducer (to calculate strain) is given by $\xi_1 + \xi_2$. Remember, at the frequency of fundamental resonance, $\lambda = 2l$ in the transducer, so $kl = \pi$. Solve for $u_f = -\omega D \sin(\omega t - k_2 l)$.

5.4. Use Equation (5.12) to solve for the equivalent electrical circuit of an air-backed transducer at resonance. Procedure: Substitute Equation (5.5) into Equation (5.12), then integrate with respect to z from $z = 0$ to $z = l$. The integral of electric field is voltage ($V = \int E_i \, dz$) and let σ_i be a constant. Solve for the total charge $q = A\sigma_i$ in terms of V and the displacements ξ of the two faces. Then, find the current $I = dq/dt$. Here, you will need to remember that $d\xi/dt = u_f$ of the faces (given by Equation (5.6) with opposite signs for the two faces) and let $V = V \cos \omega t$, so $E_i = (V \cos \omega t)/l$. The electrical admittance is finally given by I/V. Show that the admittance is the sum of two parts, one due to a capacitance with an admittance of magnitude ωC_0 (90° out of phase from V), and the other due to a resistor with admittance $1/R_m$ (in phase with V). Check your answers for C_0 and R_m with Equations (5.16) and (5.17).

5.5. An air-backed PZT transducer is radiating into water at its fundamental resonant frequency of 3 MHz. It has a surface area of 5 cm². Find its equivalent electrical circuit, including values for the components (assume it is internally lossless). When driven with a sinusoidal peak voltage of 10 V, use R_m to calculate how much power is radiated by this transducer.

5.6. Using Figure 5.9, derive Equation (5.24) for axial resolution. Find an approximate numerical value (including units) for the axial resolution of a bioinstrument whose transducer has a frequency $f_1 = 2.25$ MHz and a $Q = 5$.

5.7. (a) Find the approximate Q of the following quartz transducer arrangement at its fundamental frequency of 2 MHz by using Equation (5.18):

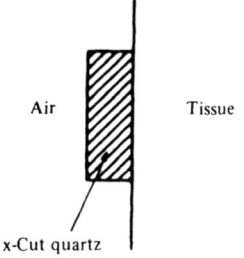

Assume that the internal losses of the transducer are zero, so that losses are due entirely to the transmission of power through the transducer faces. (*Hint:* Assume an internal wave with intensity I_0 is bouncing back and forth between the faces. Determine how much intensity is lost during one period of the fundamental frequency. Let the stored intensity be an average of before and after the bounces.)

(b) Calculate the axial resolution for this quartz transducer.

5.8. To investigate the reasons for adding an absorber to the back face of a transducer, redo Problem 5.7 for the same frequency and quartz material, but replace the air with a backing material made of an absorber whose acoustic impedance $Z = 3 \times 10^6$ kg/m² s is closer to quartz. Assume that all power radiated into the absorber is lost. Find the new Q and the axial resolution.

5.9. Redo Problem 5.7 for the same frequency and *air* backing, but use PVDF as the transducer material instead of quartz. What are the Q and the axial resolution now?

5.10. Applying the method outlined in Problem 5.7 to a general transducer with impedance Z_c radiating on one side only into a medium with impedance Z_2, show that an approximate expression for the transducer's Q at its fundamental frequency is given by

$$Q \approx \frac{Z_c \pi}{2Z_2}$$

when $Z_c \gg Z_2$.

5.11. Using geometry in Figure 5.11, show that r' is given by Equation (5.26).

5.12. Derive Equation (5.30) for the on-axis pressure field of a circular transducer starting from Equation (5.29) and following the steps outlined in the text.

5.13. Derive Equation (5.37) from Equation (5.36) using the integral relationship between J_0 and J_1 given in the text.

5.14. Integrate Equation (5.34) over a rectangular source of dimensions $b \times h$ using the geometry shown in Figure 5.16a to get the pressure radiation pattern of Equation (5.44) along the x_1 axis.

5.15. (a) Plot the pattern (similar to Figure 5.15a) of intensity measured on a plane 50 cm away from a 2-MHz unfocused circular transducer whose diameter is 1 inch. Find the diameter of the closest null ring surrounding the central peak of the pattern.
(b) Estimate the FWHM diameter of the central peak as given by the width to the -3-dB points on either side of the peak.

5.16. Find the near-field to far-field transition distance and the far-field divergence angle for each of the unfocused transducers listed below:
(a) Diam. = 1 cm, frequency = 1 MHz
(b) Diam. = 3 cm, frequency = 1 MHz
(c) Diam. = 1 cm, frequency = 2.25 MHz

5.17. An unfocused circular transducer is used in the following configuration at a frequency of 3 MHz:

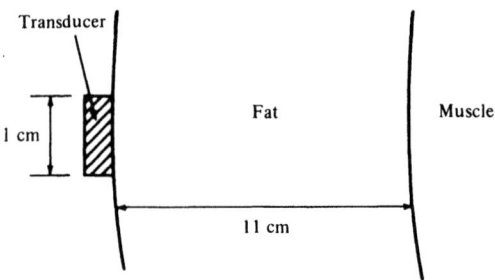

Estimate the power which the transducer must radiate into the tissue in order to receive 2×10^{-8} W back at its face from the echo due to the fat/muscle interface. Include the effects of *beam spreading* and list any simplifying assumptions you make.

5.18. Use Snell's law and ray racing to show that the focal length of a plano-concave lens is given by Equation (5.48). (*Hint:* Consider a ray entering parallel to the axis and use the small-angle approximation to find the distance where it intersects the axis.)

5.19. How large a diameter would a focused circular transducer of frequency 1.5 MHz have to be to give a focused spot size of 1 mm at a distance of 10 cm from the transducer? What would be the depth of focus of this beam?

5.20. Find the theoretical axial resolution and lateral resolution at a distance of 6 cm from a circular *unfocused* transducer whose frequency is 3 MHz, whose diameter is 1.5 cm, and whose Q is 10.

5.21. In echocardiography it is desirable to image the mitral valve leaflets with a resolution of approximately 2 mm. The distance from the chest wall to the valve is about 7 cm. To avoid excessive attenuation, a frequency of 2.25 MHz is used.
   (a) Determine the maximum Q allowed for the transducer which will give the required resolution.
   (b) Determine the minimum diameter of the lens (and therefore the transducer) which will give the required resolution, assuming focusing on the valve.

5.22. (a) Sketch a rough polar power density plot for a coherently excited linear array composed of 16 square elements, each 1 mm wide with a center-to-center spacing of 2 mm. The frequency is 2.25 MHz. Calculate the following three important features: width of main lobe; angular positions of the grating lobe(s); and ratio of peak power density in first grating lobe to peak power density in main lobe.
   (b) Explain qualitatively how each of the above three pattern features would change if each of the following modification was made independently in the array (all other parameters stay as specified):
      (i) The wavelength was decreased.
      (ii) The spacing between elements was decreased.
      (iii) The number of elements was decreased to eight.

Chapter 6
Diagnostic Imaging Configurations

6.1 INTRODUCTION

In this chapter we categorize some of the important classes of ultrasonic clinical imaging instruments and give the major operating characteristics for each category. Typical examples of each type of instrument are presented as well as areas of clinical application. No attempt has been made to be exhaustive in the coverage of clinical instruments since these devices are currently being introduced to the clinic at such a rapid rate that complete descriptions would be difficult and quickly outdated. Rather, general engineering principles relating to each category are presented.

Also, only a brief discussion is given regarding the pathological or anatomical interpretations of the images derived from these instruments. The reader interested in those topics will find several good textbooks on the clinical interpretation of ultrasonic images.

Table 6.1 outlines a commonly used classification scheme for the general configurations practiced in medical ultrasound today. A-mode, B-mode, and C-mode instruments give spatial information about the region imaged, while M-mode and continuous wave Doppler devices give motion or velocity information. The oldest technique is A-mode, which gives one-dimensional information; but except for a few specialized applications, it has been largely supplanted by the two-dimensional images of B-mode. As can be seen from Table 6.1, the category of B-scanners makes up the largest group of instruments now in use. The most sophisticated instruments (and the ones that generally give the greatest amount of data and are the most

TABLE 6.1 GENERAL CLASSIFICATION OF ULTRASONIC DIAGNOSTIC INSTRUMENTS

A-Mode (one-dimensional)
B-Mode (two-dimensional)
  Manual scanners
  Real-time scanners
    Mechanical
    Electronic
      Linear array
      Phased array
M-Mode (motion)
C-Mode (through-transmission)
Doppler (velocity)
  Continuous-wave
    Audible-output
    Analog-output
    Spectrum-output
  Doppler imagers
    CW Transverse scanners
    Pulsed Doppler
  Combined with B-mode (duplex)
Miscellaneous

expensive) are the electronically scanned B-mode imagers and the pulsed Doppler imagers. Many of the newest devices are capable of operating in a combination of modes, such as an M-mode line interlaced in a real-time B-mode sector, or the combination of pulsed Doppler and B-mode instruments (duplex scanner). In the following sections we discuss in order each of the categories shown in Table 6.1, except for Doppler devices, which are covered in detail in Chapter 7.

6.2 A-MODE

This mode, like all other imaging configurations except C-mode and continuous Doppler, is based upon the *pulse echo* technique, wherein a short pulse of ultrasound is transmitted by a low-Q transducer into the tissue regions being investigated. Reflections from each of the various tissue boundaries due to changes in acoustical impedance are received back at the transducer, and the total transit time from initial pulse transmission to reception of the echo is proportional to the depth of the boundary. This makes possible a one-dimensional mapping of the tissue interfaces along the line of propagation of the beam.

6.2 A-MODE

What distinguishes A-mode from other modes is the method of display. The "A" can be considered to stand for "amplitude" display. Figure 6.1 is a general schematic showing the elements of an A-mode instrument. A pulser circuit transmits a sharp voltage spike to the transducer, which emits a short pulse of ultrasound into the tissue (a simple model is shown) as described in Chapter 5. The transmit/receive (T/R) switch isolates the high-gain first-stage receiver amplifier from the large transmit voltage spike to avoid overload, saturation, or even burnout of this sensitive stage; the switch does allow, however, the low-level received echo signals to pass to the receiver. Often, a pair of parallel, reversed diodes which shunt the receiver input to ground for the large transmit voltage but which act as an open circuit for the low-voltage echoes are used for this purpose, as shown in Figure 6.2.

At the time of transmit, the pulser also triggers a sawtooth voltage generator which begins a uniform horizontal time sweep of the beam across the cathode ray tube (CRT) display. On the vertical axis of the display are presented the received signals after significant amplification and usually some signal conditioning. At the very beginning of the sweep, as shown at point a on Figure 6.1, the transducer is still reverberating from the initial pulse and is also receiving strong echoes from close surfaces; this large initial received disturbance is known as the "initial bang." This large response can be attenuated or even blanked from the display, but it serves a useful purpose in marking the front surface of the body.

As the transmitted pulse progresses through the tissue of impedance Z_1 toward the interface at depth l_1 with the organ of impedance Z_2, essentially no reflection takes place as long as the impedance Z_1 is more or less homogeneous (except for low-level scattering). The first significant signal received is the reflection from the anterior organ boundary after the echo has returned to the transducer, that is, after it has traveled a total distance of $2l_1$. As shown in more detail in Figure 6.3, the time from the initial bang to the time of arrival of this echo, denoted at point b, can be calculated as

$$t_1 = \frac{2}{c_1} l_1 \tag{6.1}$$

where c_1 is the phase velocity of the first tissue. The next echo received, with somewhat lower amplitude due to the increased attenuation of the longer overall path and partial reflection both ways at the first interface, is the echo from the posterior organ wall, denoted as point c. This signal arrives at time t_2 given by

$$t_2 = t_1 + \frac{2}{c_2} l_2 = \frac{2}{c_1} l_1 + \frac{2}{c_2} l_2 \tag{6.2}$$

where c_2 is the phase velocity in the second tissue (organ).

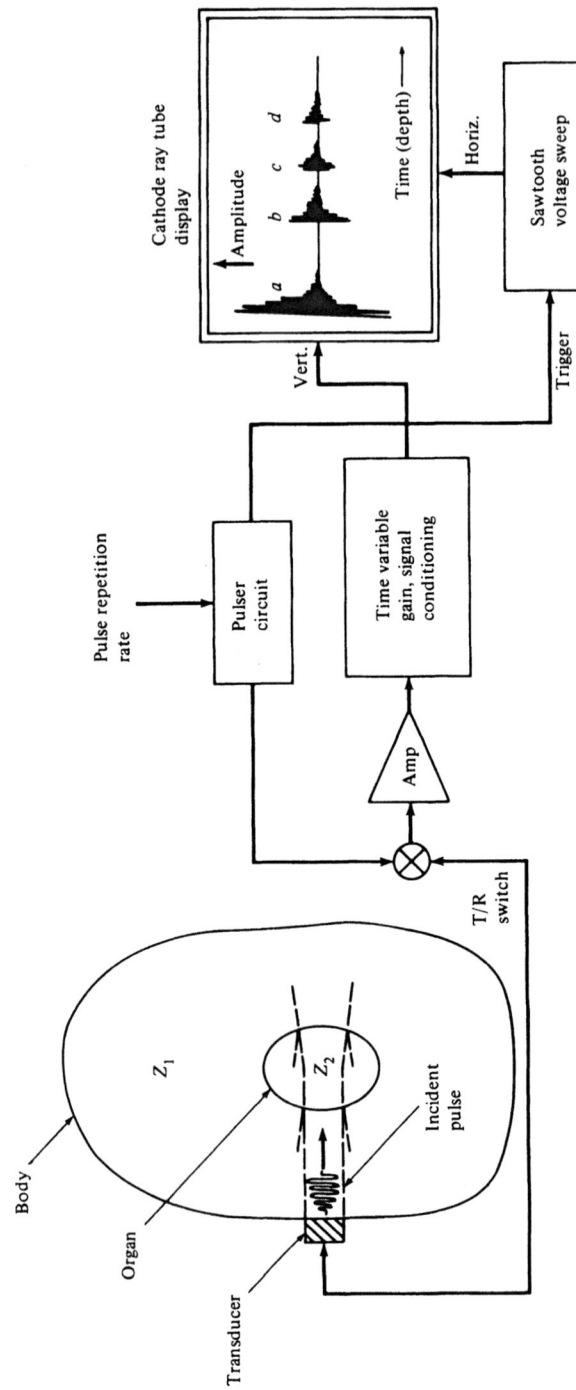

Figure 6.1 Elements of an A-mode pulse echo instrument. In the A-mode, the amplitudes of the received echo signals are presented on the vertical axis of the display; the horizontal sweep in time may be converted to equivalent depth of penetration. Signals denoted a, b, c, and d on the display are discussed in the text and in Figure 6.3.

6.2 A-MODE

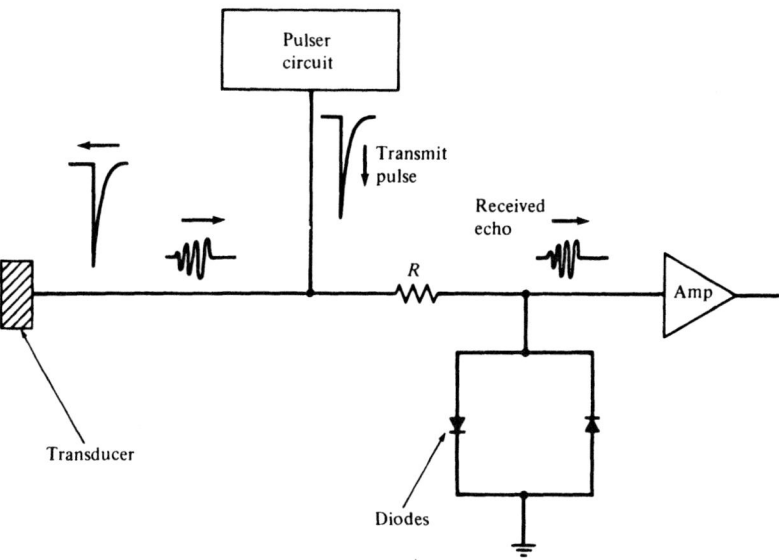

Figure 6.2 A pair of parallel, reversed diodes acts as a passive T/R amplifier protection switch, since the diodes appear as essentially short circuits for amplifier input voltages above approximately 0.7 V (such as the large transmit spike), but as open circuits for the low-voltage received echo waveforms. To avoid wasting excessive transmit power in the resistor R, the value of R should be somewhat larger than the electrical impedance of the transducer. Also, to pass the maximum received signal to the amplifier, the off-output impedance of the pulser circuit should be larger than R in series with the amplifier's input impedance, which itself should be larger than R. This leads to the requirement of a large off-output impedance for the pulser circuit.

It can be seen that to relate the time base of the horizontal sweep to the depths of the tissue boundaries, the phase velocities in each medium must in theory be known. However, it is usually difficult to ascertain a priori the exact tissue velocities, so for purposes of calibrating the tissue/depth relationship, *most American-built instruments assume that the tissue phase velocities have a value partway between those of water and muscle; that is, $c \approx 1540$ m/s.* Section 4.5 shows that this assumption produces only a small error for soft tissues. Thus, the proportionality constant relating distance to time is given by

$$\boxed{\frac{2}{c} = 1.3 \times 10^{-5} \quad \text{s/cm}} \qquad (6.3)$$

which is equivalent to 13 μs for each 1 cm of depth (round trip) traveled.

The signal at point d is an artifact due to *multiple reflections;* in this

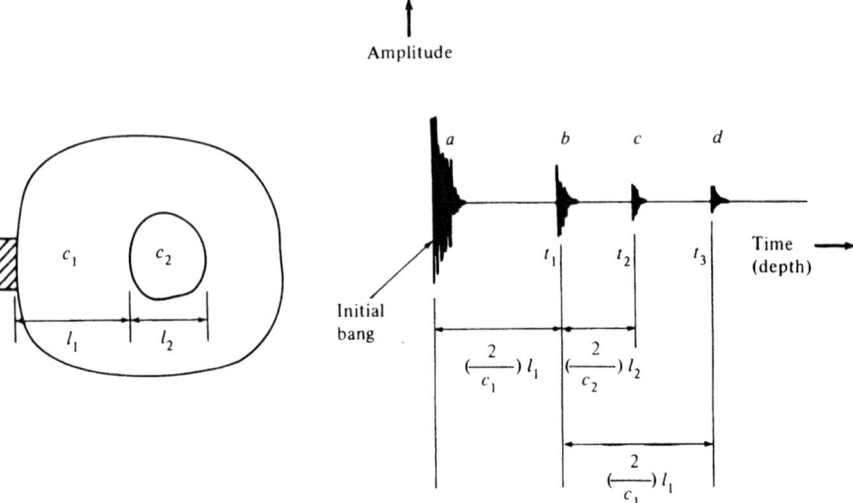

Figure 6.3 Details of the arrival times of echoes from the interior tissue boundaries. The boundary depths can be related to echo arrival times through the proportionality constant $2/c$.

case, a portion of the first echo arriving at the transducer face is reflected back into the tissue, where it is again reflected from the anterior organ boundary back to the transducer at time t_3. Note that $t_3 = 2t_1$, so one clue for distinguishing false boundaries due to multiple reflections is the uniform spacing of the echoes. While later multiple reflections will be low in amplitude because of the long multiple path length and will eventually fall below noise level, multiple reflections from near structures may overlap the first echoes from important boundaries, and this interference can cause problems in interpreting the images (see Problem 6.2).

The pulse repetition rate applied to the transmit pulser circuit is limited only by the requirement that almost all of the possible reflections from the previous pulse die out before the next pulse is transmitted to avoid overlap and confusion. This depends upon the depth being imaged. For imaging in the thorax, for example, where 20 cm is usually the maximum primary depth of penetration needed, echoes, including multiple ones, may be considered gone after two times the arrival time of the last primary echo, or

$$t = \frac{(2)(2)(20 \text{ cm})}{1.54 \times 10^5 \text{ cm/s}} = 0.52 \text{ ms}$$

Thus, the maximum pulse repetition rate is approximately 2 kHz. For real-time B-scans, described in a later section, it is desirable to pulse at an even faster rate to quickly fill in the image, so these scanners often provide an operator-selected tradeoff between depth of penetration and pulse repetition rate.

6.2.1 Signal Conditioning

The electronics of the receiver system tailors the received signals for maximum readability on the display screen. The first-stage amplifier must possess high gain, low noise, and sufficient frequency response to amplify the incoming echoes. The next stage provides further gain, and it is advantageous here to make this gain vary according to the elapsed time since the initiation of the transmit pulse. This *time gain control* (*TGC*, also known as swept gain, or time gain compensation) helps compensate for the ever-decreasing signal strengths from deeper tissues due to the greater attenuation over longer paths. Thus, the gain ramps up at a rate that tends to balance tissue absorption. Figure 6.4a shows a simple example of a linearly increasing gain. The exact attenuation encountered versus depth depends upon the tissue types encountered, but Section 4.5 shows that a typical attenuation for soft tissue is 1 dB/cm MHz. Since attenuation expressed in dB is additive over successive path lengths, the rate of *TGC* increase should be approximately 1 dB/MHz for each interval of time that corresponds to 1 cm of pulse travel. Assuming $c = 1.54 \times 10^5$ cm/s, the gain should therefore increase at a rate of

$$TGC \text{ Rate} \approx +154 \quad \text{dB/ms} \tag{6.4}$$

for each MHz of the transmitted frequency (see Problem 6.3).

Actually, the optimum rate is somewhat lower than this, due to two factors. First, the transmitted pulse is not single frequency but is composed of a broad spectrum of frequency components as determined by the Fourier transform of the pulse's time waveform. The exact shape of the frequency spectrum of the pulse depends upon its time envelope, but in general terms it can be said that a pulse of fundamental frequency f_1 whose envelope has a time duration of t' will possess frequency components that stretch an amount

$$\Delta f \approx \frac{1}{t'}$$

in width, centered around frequency f_1.

For example, Figure 6.5 shows the frequency spectrum for the simple case of a single rectangularly shaped pulse of frequency f_1. If the frequency spread of the spectrum is taken as half the distance between the first zeros on either side of the peak, then $\Delta f = 1/t'$ (see Problem 6.4). Thus, the shorter the pulse, the broader is the frequency spectrum contained in the

(a)

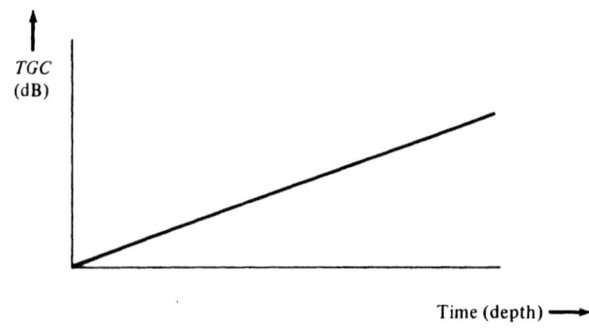

(b)

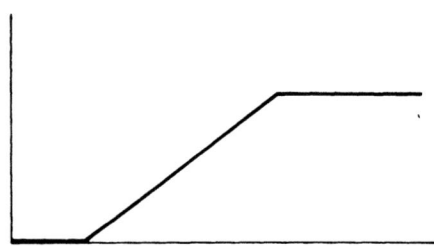

(c)

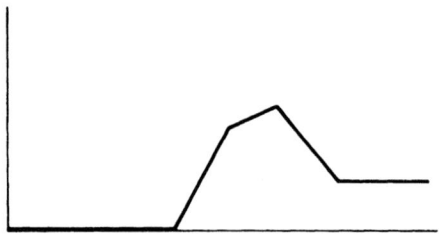

Figure 6.4 Some curves of different time gain control (*TGC*) as a function of depth to compensate for tissue loss. (a) A monotonically increasing gain curve may be theoretically optimum but is little used in practice. (b) Often it is desirable to deemphasize near boundaries and to avoid high amplification of noise at very deep boundaries, as this curve accomplishes. (c) If a certain structure at a known distance is to be highlighted, such as a heart valve, this "pedestal" gain shape is appropriate.

6.2 A-MODE

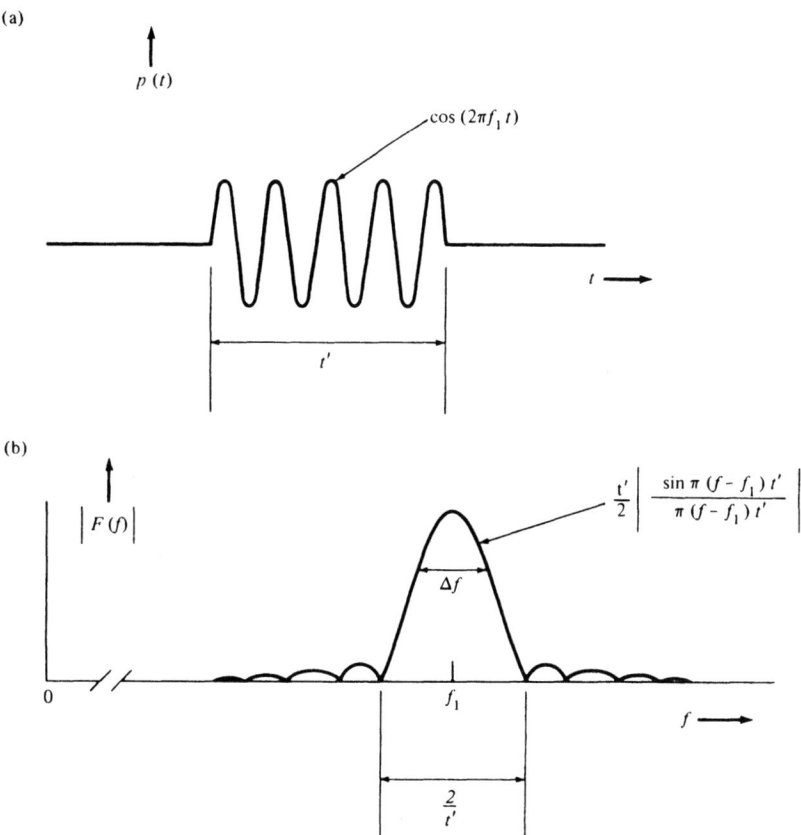

Figure 6.5 (a) A rectangular waveform model for a single transmitted pressure pulse. (b) The spectrum of the pressure in the frequency domain obtained by the Fourier transformation of the pulse in (a) (see Problem 6.4); the spectrum has the form (sin x)/x. The spectrum will be continuous strictly only for an isolated pulse but is approximately correct for the low-duty cycle pulse trains used in medical imaging.

pulse. The frequency components higher than f_1 are attenuated the fastest because of the dispersive absorption nature of tissue discussed in Chapter 4, leaving the components lower than f_1 to carry the bulk of the pulse's energy after it has traveled a reasonable distance. These lower-frequency components have lower attenuation than predicted for frequency f_1, and therefore less *TGC* boost is needed.

Second, the tissues that make up the path of the ultrasound pulse are often less attenuating than 1 dB/cm MHz, particularly for fat layers, blood in the heart vessels and chambers, and the fluid surrounding the fetus. Thus, less *TGC* is required than given by Equation (6.4).

In fact, on most modern clinical instruments, the rate of *TGC* is variable and under the control of the operator to meet the specific needs of the application. The penetration depth is divided into several segments, and a separate sliding knob determines the rate of *TGC* increase for each segment. Often, the selected gain-versus-depth curve will be displayed on a portion of the cathode ray tube for visual checking. This flexibility allows tailor-made gain curves such as shown in Figure 6.4b. It may be desirable to avoid giving too much attention to structures close to the transducer, so the gains of the initial segments are set constant and low. Echoes from boundaries much deeper than the region of interest may also be deemphasized by setting the gain constant there, especially since the very high gain needed for deep echoes may be amplifying an appreciable amount of noise, making interpretation of this part of the display confusing. Figure 6.4c shows how a specific structure may be highlighted by increasing the gain just in the neighborhood of the structure, giving a window, or "pedestal"-shaped, gain curve; this effect is sometimes used in cardiac imaging.

The receiver electronics may perform further signal conditioning tasks. The voltage waveforms produced by the transducer in response to the received pressure pulses will follow the same shape as the pressure waveform, with positive and negative excursions about zero voltage. This is because the voltage output of a piezoelectric transducer is proportional to applied pressure, as shown in Chapter 5. Often these input pulses are electronically *demodulated* by a diode stage followed by a capacitor such that only the envelope of the pulse, not its oscillation at the ultrasonic frequency f_1, is amplified and displayed. This destroys information regarding the phase and exact timing of the received pulses, but makes the display easier to read, especially in the B-mode. Since it essentially removes the carrier frequency f_1, demodulation will shift the frequency spectrum down such that it is centered around zero frequency as shown in Figure 6.6. This lowers the frequency response requirements on subsequent electronic stages. Some instruments allow control of the time constant of the demodulation and thus of the processed and displayed pulse envelopes. It should be remembered, however, that any lengthening of the pulse envelope such as caused by a limited frequency response of the demodulation or other electronic stages will degrade the axial resolution of the system. Fortunately, axial resolution is normally much better than lateral resolution, and there is therefore some room for tradeoff between axial resolution and the speed of the electronic circuits.

Due to the difficulty of presenting a wide variation of signal strengths adequately on the cathode ray tube, the wide dynamic range of the detected echoes may be *compressed* to a smaller range by a logarithmic amplifier stage before display. This allows small echoes to be seen on the same screen as large echoes for a given gain setting. However, very small echoes (and the noise itself) are disproportionately amplified with this technique, so

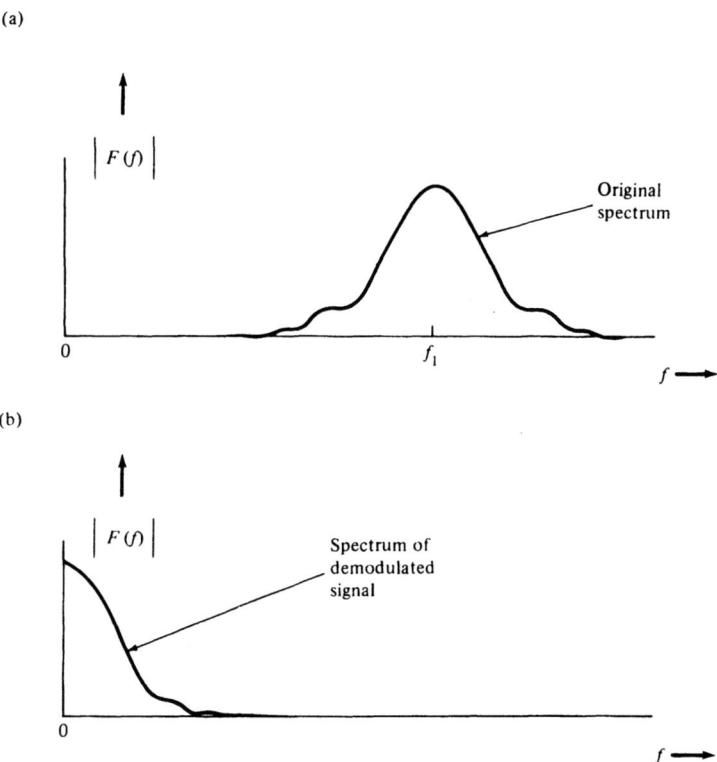

Figure 6.6 Performing demodulation on the signal whose original frequency spectrum is shown in (a) will shift it down to a center around zero frequency, as shown in (b).

many instruments have a variable *threshold* feature to blank from the display all signals whose amplitude is below the setting of the threshold value. This may be used to eliminate noise or other inconsequential receiver signals from notice. More will be said about dynamic range in the next section.

To summarize, Figure 6.7 shows how a series of echoes would appear as they are processed by various stages of the signal-conditioning electronics. The demodulated A-mode display would look like Figure 6.7e on the screen.

6.2.2 Applications of A-Mode

The advantage of the A-mode is that it gives positional information quickly with a minimum of sophisticated equipment. Its weakness is that this depth information is one-dimensional only, along the line of beam propagation;

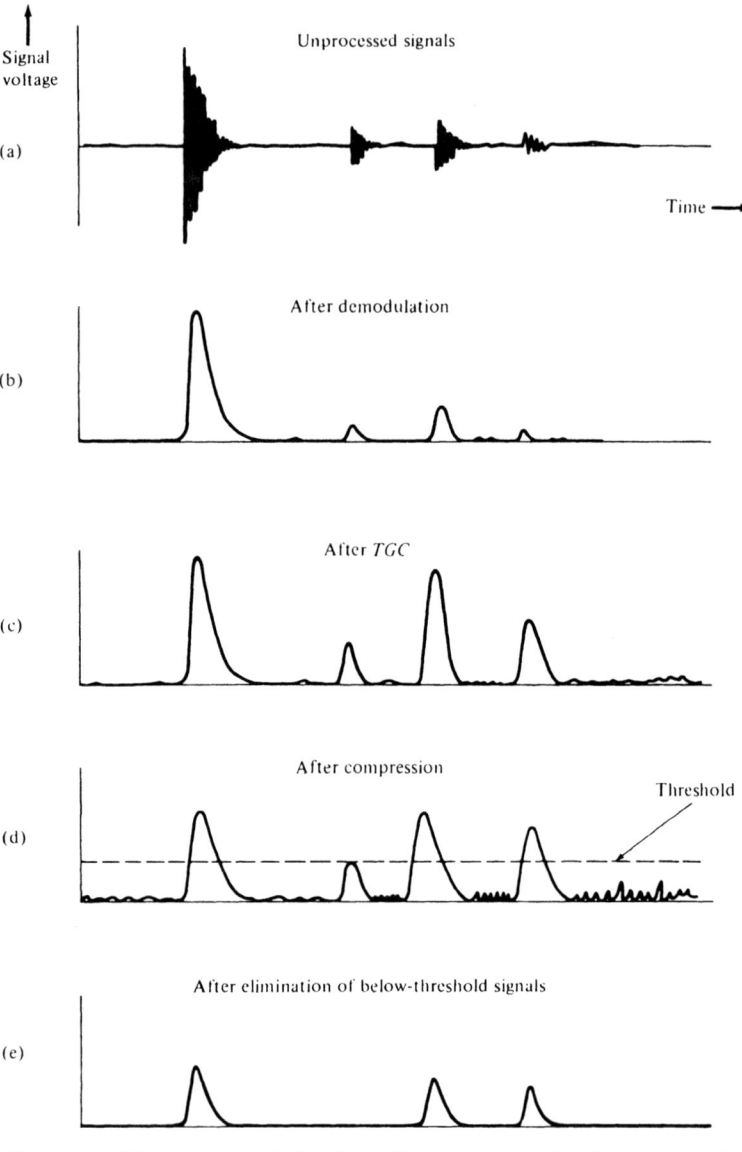

Figure 6.7 The sequence of signal conditioning steps often implemented in processing the received ultrasonic echoes: (a) unprocessed signals after first stage of amplification; (b) after demodulation, yielding pulse envelopes; (c) time gain control (*TGC*) amplification; (d) logarithmic compression; (e) elimination of signals below threshold setting.

the operator's imagination must be used to picture two-dimensional images from the A-mode display.

A-Mode has been used extensively in midline echoencephalography, where the position of the midline features of the brain is determined in relation to reference echoes from the near and far skull boundaries. In the healthy brain, the midline structures will be positioned at the center of the skull in the median sagittal plane. If the structures are moved significantly to one side or the other of this plane, however, by an expanding lesion such as a brain tumor or hemorrhage, the asymmetry may be easily recognized on the A-mode display. Image quality is not optimum due to the intervening skull bone through which the ultrasound pulse must pass; but by careful transducer placement, signal strength is sufficient to determine symmetries.

Ophthalmological scanning is also successfully carried out in the A-mode. When the eye is to be probed with ultrasound, the small size of the A-mode transducer compared to the somewhat larger heads required by other modes is an obvious advantage. Measurements of eye size and growth patterns are accomplished by this technique, as well as the detection of tumors or other pathology and the location of foreign objects (such as metal fragments) for surgical removal. The short penetration depths required allow the use of high frequencies, from 5 to 15 MHz, for excellent resolution.

Figure 6.8 shows a photograph of several styles of A-mode probes intended for ophthalmological use, along with a typical display.

6.3 B-MODE

The development of this mode can best be understood by considering one line of an A-mode display that is modified so that the amplitude of the returned signal does not cause a vertical displacement of the sweeping beam of the cathode ray tube, but rather causes a corresponding increase or decrease in the *brightness* of the beam as seen on the screen. B-Mode stands for "brightness" modulation (z-axis modulation) of the displayed beam. Figure 6.9 shows how one line of a B-mode is obtained corresponding to the A-mode display of the same object.

Brightness modulation frees one axis of the display to be used for presenting other information. The axis along the direction of the beam still corresponds to depth of penetration or distance; but for the B-mode the axis perpendicular to the beam is made to display distance also by linking the direction of writing of the beams on the display to the actual direction of the ultrasonic beam propagating from the transducer. This is usually

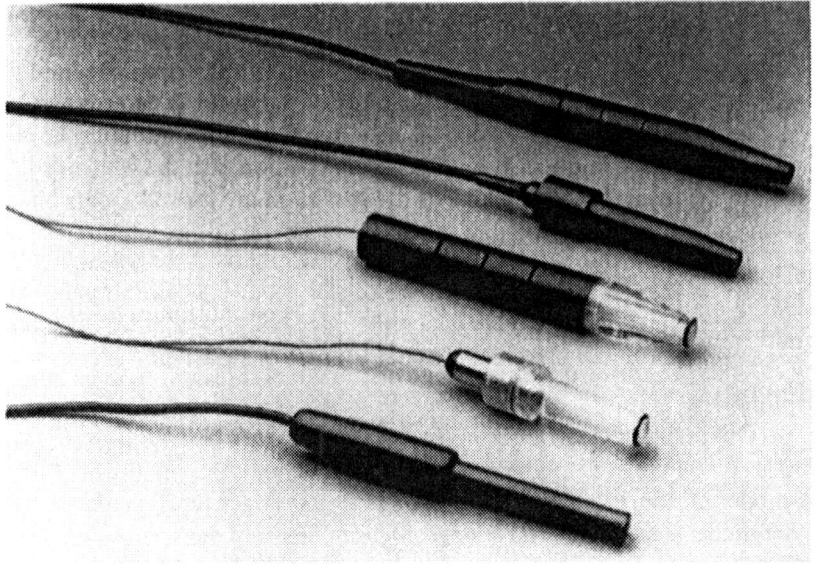

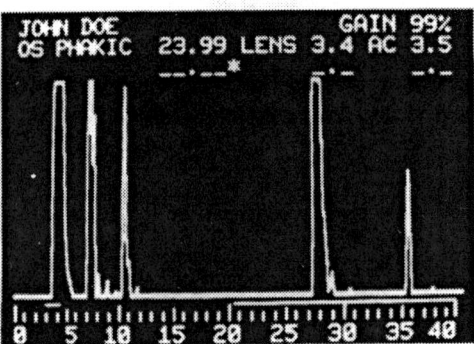

Figure 6.8 (a) A family of probes used for A-scanning of the eye. The stand-off path between transducer and eye lid may be solid or water-filled. (b) A typical A-mode display for measuring distances of ocular structures along the eye axis. B-mode often supplements the exam. Photographs courtesy of CooperVision Surgical.

done with electrical position-measurement devices called position transducers. These devices measure the angle and displacement of the ultrasonic transducer head with respect to a fixed frame of reference (for example, the massive cabinet of the instrument) by means of mechanical arms connected to the head. The mechanical linkage is cleverly designed to allow very flexible movement of the transducer as it is placed and manually

6.3 B-MODE

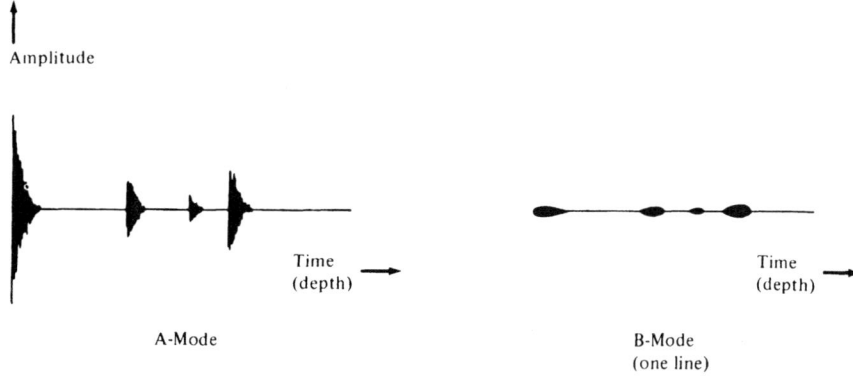

Figure 6.9 In the A-mode, the amplitude of the received signals deflects the display beam vertically. In the B-mode, the brightness of the beam is modulated by this amplitude, so each line is thin.

rotated on different surfaces of the patient. The electrical position transducers are usually linear or rotary potentiometers used as variable-voltage dividers to provide electrical signals to the beam steering electronics of the display.

Figure 6.10 shows the elements of a manually scanned B-mode instrument. The pulser circuit and receiving electronics are similar to those explained in the previous section (with *TGC*, compression, etc.), but the receiver output now modulates the brightness of each line on the display. The direction of each line (given by the angle θ) is determined by the beam steering electronics to correspond to the ultrasonic beam direction indicated by the position transducers attached to the mechanical linkage. The operator places the transducer head at the selected site on the surface of the patient and rocks the transducer back and forth, sweeping the ultrasound beam in a pie-shaped, or sector, format. The corresponding lines of the electron beam will form the same sector shape on the display, with dots of brightness revealing the positions of reflecting surfaces. The screen possesses a long persistence time (storage scopes are usually used), so that the display of all previously written lines will remain until the sector is erased. The operator watches the display while controlling the movement of the transducer head, so portions of the sector that may need more detail are filled in by dwelling on those portions. In this fashion the operator "paints" a sector on the display.

The image produced in B-mode is a two-dimensional view or cross section of the region being scanned. The beam steering module generates vertical and horizontal deflection voltages for the display which are coordinated such that the display has the same scale factor in the vertical direction as it does in the horizontal. Thus, there is no distortion of angles

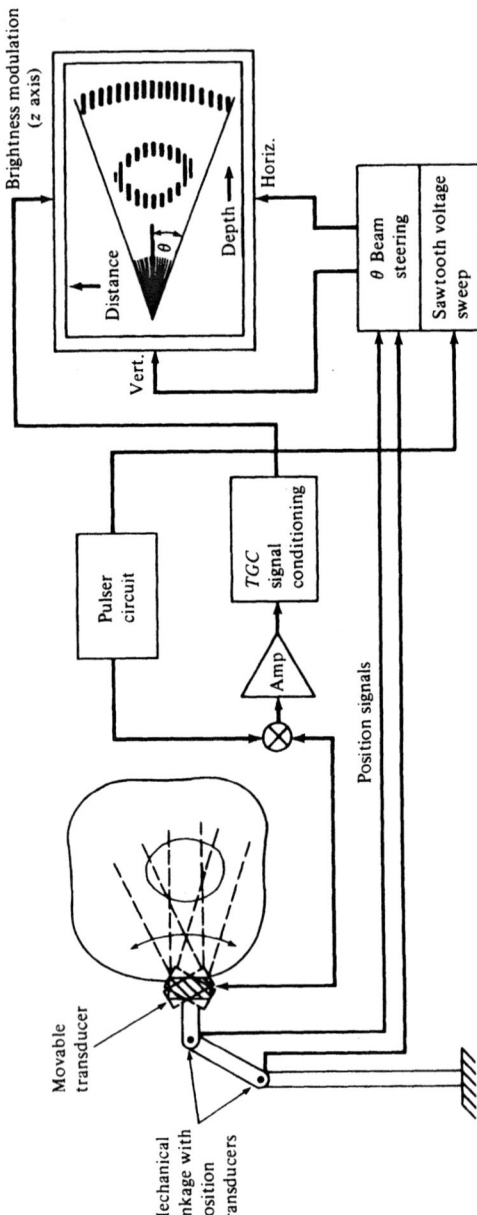

Figure 6.10 A hand-scanned B-mode instrument in which the transducer is rocked back and forth manually to produce a corresponding sector on the cathode ray tube display. The bright dots along each line of the sector correspond to the amplitude of the received echoes when the transducer is pointed in that direction; the display is a two-dimensional mapping of reflections, or "tomogram."

6.3 B-MODE

or shapes of surfaces when viewed on the screen. Since the view presented is as though it were a "slice" or cut through the body, the image is known as a *tomographic* scan ("tomo" = cut). In X-ray imaging, the view of a computerized tomographic (CT) scanner is in a similar plane, and the acceptance and familiarity of CT scan images in the modern clinic has helped the interpretation of ultrasound B-scan images. Conventional anterior–posterior X-ray films, however, are integrated views through the body and do not bear correspondence to the planes of the ultrasonic B-scan image (see Figure 1.1).

The brightness of the dots on the display can be presented in a broad range of values in modern B-scan instruments. This is known as *gray-scale* imaging, as opposed to the bilevel (bright or dark only) displays of earlier devices. Such gray-scale images present a much richer display of the range of the echo strengths, thereby making interpretation easier. The dynamic range of a typical cathode ray tube, determined by the number of separately resolvable gray-scale levels, is rather limited, about 20 dB. On the other hand, the range of echo strengths, from strongest to weakest, from a typical region being viewed (say, from the adult heart to a depth of 15 cm) is about 110 dB. The strongest echo is limited by the amount of power that can safely be generated in the transmit pulse, while the weakest echo is determined by the noise levels in the receiver system (typically several microvolts equivalent at the input of the first-stage amplifier). This 110 dB of input dynamic range will be reduced by an amount of approximately 60 dB by the action of the time gain control stage, which boosts the weak signals coming from depth as described earlier.

This leaves a dynamic range of 50 dB which must be further compressed by the signal-conditioning electronics to be adequately displayed within the 20-dB dynamic range of the gray-scale screen. The logarithmic compression stage described in the previous section will accomplish this task. Figure 6.11 shows how the dynamic range of the signals is processed from the input through to the display.

Permanent recording of the B-mode display may be accomplished by either of two techniques: (1) a record can be obtained directly from the screen by a film camera or by a video camera with an associated video tape recorder; or (2) a scan converter module may be used to change the sector-shaped polar coordinate pattern of signals coming from the receiver into a raster format for direct electrical storage on a video tape recorder, avoiding the optical link of the previous technique. In fact, with a scan converter the signals may be displayed on a video monitor as the primary display device rather than an x–y CRT display. Either way, an accurate and convenient means of permanently recording selected scans from each patient is mandatory in today's clinical environment.

If a digital scan converter is used, storage in electronic memory will also allow some postreception processing to be done, such as spatial av-

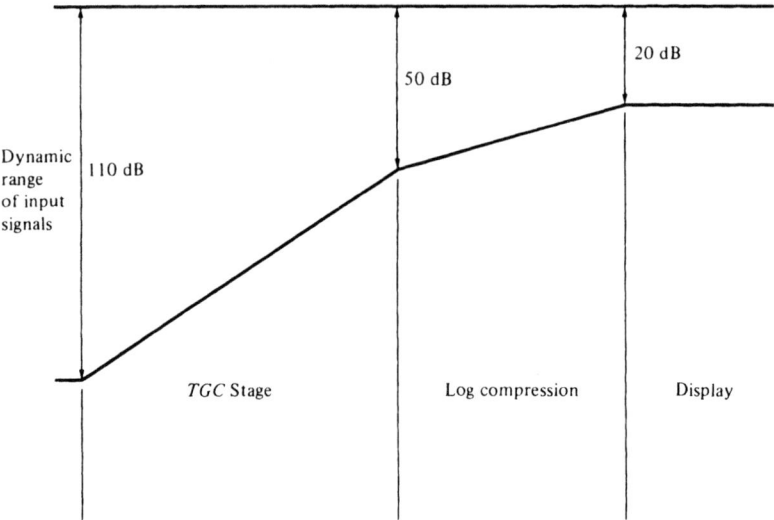

Figure 6.11 The dynamic range of the input signals is reduced by the time gain control stage (by approximately 60 dB) and by logarithmic compression to match the 20-dB capability of the gray-scale display.

eraging to reduce speckle noise or modifying the linearity of the gray-scale conversion to emphasize certain amplitude regions.

6.3.1 Compound B-Scanners

One difficulty with B-mode scanning from a single incident direction is that any flat reflecting interface which is not nearly perpendicular to the beam's line of propagation will not return much power to the transducer since the law of specular reflection dictates that the reflected beam will bounce off at an angle and not be collected by the transducer. Although most interior body surfaces are not really planar and irregularities of these surfaces will scatter some of the incident beam into a broad pattern, it is still very desirable to be able to illuminate the object from several different directions; what one angle misses may be seen from another.

Therefore, many manually scanned B-mode instruments have the capability to *compound* their images; that is, their mechanical linkages allow the transducer to be moved and placed at different locations on the patient's surface, and the sector displays obtained from each location are correctly superimposed on the display, thereby filling in detail of the object from various directions. This feature is a powerful aid in analyzing complicated structues as the operator views the developing images on the screen.

To be accurate, a compound B-scan instrument must minimize misregistration error caused when the sector obtained from one direction is not accurately spatially oriented on the display with respect to an overlapping sector obtained from another direction. One major cause of misregistration is inaccuracy, wobble, or looseness in the position transducers of the mechanical arm when locating the transducer head in space. When registering compounded sectors, these elements must meet rigid accuracy specifications.

Another cause of misregistration is inherent in the inhomogeneities of the tissue region itself. The proportionality constant $(2/c)$ relating time to distance on the display is assumed uniform and constant throughout the imaged region. In reality, the phase velocity will vary from path to path depending upon tissue type, and if the intervening tissue encountered when viewing an object from one direction has a different velocity than that encountered when viewing from another direction, the two overlapping images of the same object will not be completely registered under the assumption of a uniform c (see Problem 6.7). To the author's knowledge, no current clinical instrument accounts for this source of error.

6.3.2 Applications of B-Mode

B-Mode instruments represent the vast majority of devices used in the clinic today. This is due to the great variety of anatomical regions that may be successfully scanned with B-mode (both manual scan and real-time) and to the easy interpretation of the two-dimensional tissue maps characteristic of the B-mode.

One prime application is in obstetrical cases, where fetal growth rate, orientation, and abnormalities may be charted without the risk to the fetus attendant with X-ray imaging. Placenta location or possible multiple fetuses may also be verified. If amniotic fluid samples are to be drawn, ultrasound views will help in proper insertion of the needle. In gynecological scans, suspected ovarian cysts or malignant tumors may be investigated in the B-mode. Chapter 8 discusses the safety aspects of using ultrasound for fetal scanning. Figure 6.12 shows a typical manually scanned compound B-mode instrument along with an image of a pregnant uterus.

For abdominal scanning, B-mode often gives clear images of the liver

(a)

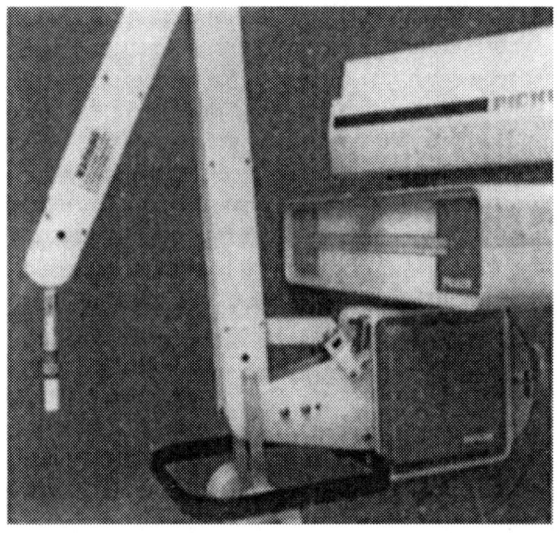

(b)

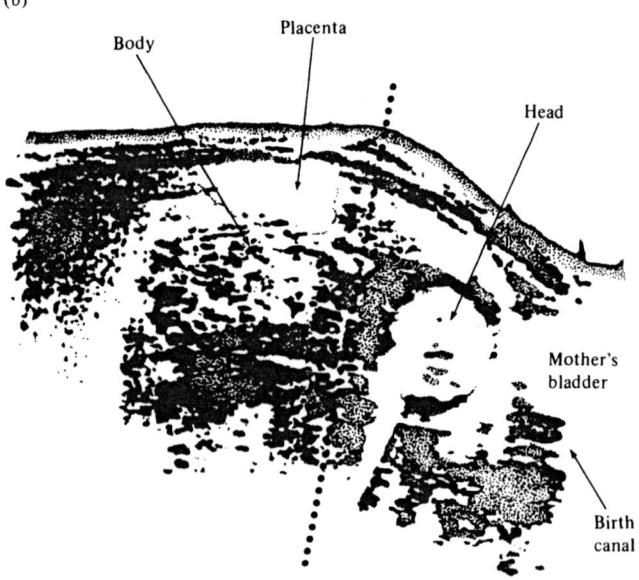

Figure 6.12 (a) A typical manually scanned compound B-scanner. (b) A compound B-scan of a pregnant uterus, showing fetal features.

(to assess diffuse liver disease or malignant tumors), the spleen, the gallbladder, and the kidney. Abnormalities due to anatomical changes from tumors or other lesions are apparent. Often frequencies as high as 5 MHz are used here to increase resolution for short penetration depths.

The female breast is another region that is successfully scanned by ultrasound. The presence of cysts or tumors may be detected by this relatively convenient and safe procedure. The soft tissues of the breast provide an ideal medium for ultrasonic propagation.

Imaging of the heart and its movement during a cardiac cycle presents special challenges to manually scanned B-mode instruments. Since the heart is surrounded by lung tissue with air-filled alveoli and by the bony cage of the ribs, sonic access to the heart is limited to a few "cardiac windows" where neither air nor bone intervenes, such as at the narrow intercostal spaces on the anterior chest wall. Even more important, movement of the heart walls and valves occurs at a much faster rate than a B-mode sector can be obtained by manual rocking, so images of these structures will be blurred on a normal hand-scanned instrument. This has spurred development of real-time B-scanners whose sector sweeping is done mechanically or electronically in such a short time that the image appears to be essentially real time; these instruments are discussed next.

6.4 REAL-TIME B-SCANNERS

One of the operational advantages of manual scanning is that it allows the user to tailor the scanning motion to emphasize structures of interest, concentrating on certain features for more resolution or viewing from several different directions to fill in patterns of irregular angular reflection. The main disadvantage of manual scanning, however, is that a sector cannot be completed in the short time needed to freeze motion associated with the heart. Since a typical heart cycle is less than 1 second in duration, one complete sector scan should be obtained in a fraction of a second to avoid blurring the heart details (such as valves and wall position).

Fortunately, there are several ways to achieve this. They fall into two broad categories: mechanical scanning and electronically facilitated scanning. The sweep rates possible with these devices range from a few sectors per second to as high as several hundred sectors per second, depending upon the method employed, and the depth, width, and line density chosen for the sector. Thus, the sectors are obtained fast enough to warrant the name "real time" to differentiate from the relatively slow manual technique.

In addition to the advantage of quickly capturing a heart image, real-time scanning has other benefits. There is no need for the complicated mechanical–electrical linkages to provide spatial orientation information to the display since referencing of the sector lines to each other is inherent in the sweeping mechanism. Also, searching the target area for an anatomical landmark or a particular feature is greatly speeded up due to the fast acquisition time and almost continuous display of successive sectors.

This quasi-continuous display of real-time scanners leads to a certain perceptual advantage also, apparently due to a phenomenon in the workings of the human visual system. When presented in a motion picture fashion, the structures being viewed—whose boundaries often have a mottled appearance from the effects of interference (speckle), irregular reflectivity, and noise—will tend to be perceived more connected and "whole" than when presented in a static single-frame manner. This effect is especially evident when the viewer is trying to discern moving components such as vessels, heart walls, and valves.

The price paid for these benefits is an increased transducer head volume (since the sweeping motion is performed by mechanical parts in the head or by multiple elements in the array) and a higher system cost. Nevertheless, real-time scanners are essential for cardiac viewing and are also popular for abdominal scans.

6.4.1 Mechanical Scanners

Fast sweeping of the beam direction can be accomplished in a variety of mechanical ways. Figure 6.13 shows three possibilities. The transducer can be mounted on the edge of a rotating wheel such that its beam radiates through an opening in the housing whenever it passes by. To reduce the dead time of the system, which occurs during each rotation when the transducer is not positioned somewhere within the window, more than one transducer can be arranged around the wheel's periphery. Rotary motion via an electric motor is relatively fast and is smoother than oscillatory motion, but electrical contact to the transducer(s) must be made through rotating slip rings, which are electrically noisy, or by a complex transformer arrangement.

To decrease the dead time even more, a single transducer can be made to oscillate back and forth in a manner characteristic of manual rocking, but faster. The beam's path sweeps through the desired sector extent. This technique allows the beam to be useful at all times and simplifies connection to the transducer's electrical terminals, but the cyclical stop-and-start motion of the rocking limits the sweeping speed to a few sectors per second.

Decreasing the mass of the oscillating part helps reduce the momentum change needed each cycle, thereby reducing vibration, so sometimes

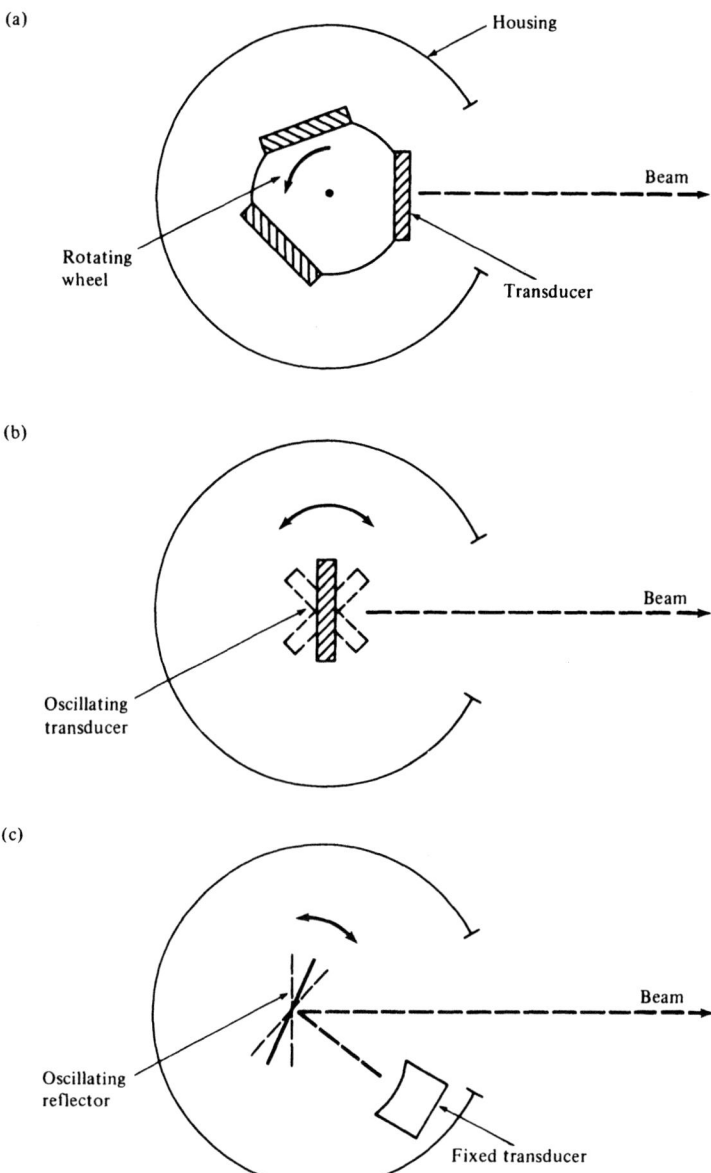

Figure 6.13 Some ways of mechanically sweeping the beam to obtain B-scan sectors: (a) rotating the transducer(s), (b) wobbling the transducer, (c) wobbling a reflector with the transducer fixed.

a lightweight sonic mirror is oscillated instead, reflecting the beam from a stationary transducer into the sector pattern.

All of these techniques have compromises stemming from their mechanical nature. The heads of the transducers are heavier and more bulky than nonswept transducers. The problem can be decreased by employing smaller transducer diameters, but then lateral resolution is worsened. Also, there may be a problem with multiple echoes occurring inside the housing due to reflections from the window and other parts, leading to an inability to image structures close to the transducer window.

Electronic scanning overcomes many of these objections, but at a higher system cost because of the complex electronic circuitry required to accomplish the beam steering.

6.4.2 Electronic Scanners

When the beam is swept electronically, the motion can be obviously faster and more flexible with no mechanical vibration or constraints to worry about. There are two main ways to achieve this; both use arrays, but they differ in whether the elements are excited one at a time or coherently with a certain phase relationship. The concept of an array has already been introduced in Chapter 5. The beam patterns from multiple-element transducers were discussed and plotted in Section 5.6.

Linear Array

In this approach the array elements are pulsed sequentially to produce a rectangular scan pattern (not a pie-shaped sector). At any time only one element (or a small group of elements) is in operation. Figure 6.14 shows the basic configuration. As each element is pulsed, its beam is more or less confined to radiate straight ahead in a corridor centered at the position of the element in the array. The element receives echoes from its transmitted beam before the next element is excited. The received echo strengths are displayed in B-mode fashion on the screen; the scan line position corresponds to the position of the element in the array.

As is common with electronic scanning, the entire array can be sequenced in a very short time, even though the array may contain many elements (typically 64 or more) to cover a large viewing area. For example, if the pulse repetition rate is 2 kHz, a 64-element linear array will be scanned in 32 ms.

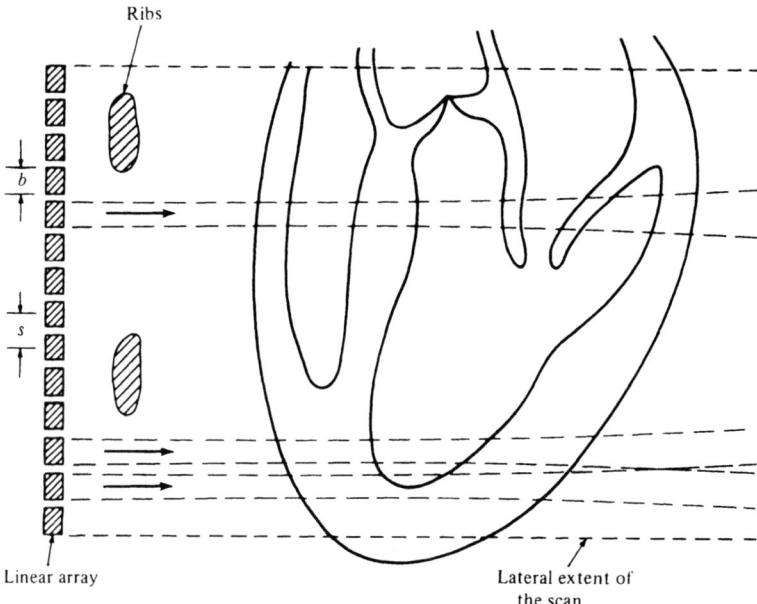

Figure 6.14 Schematic of the beam directions and display format from a linear-array real-time scanner, here used to image the heart. Only the beams from three of the elements are shown for simplicity. The elements are generally operated one at a time. The scanned area is displayed as a rectangular cross section on the viewing screen.

Characteristics and Advantages of the Linear Array

The elements do not have to be excited in regular order, starting at one end and progressing through each element in turn to the other end. In fact, there are good reasons not to do this. First, by jumping the order of firing around within the array in a carefully determined pattern, the display can be refreshed in a more uniform manner, avoiding any "windshield wiper" effect. Also, it is advantageous to let the deep echoes die out completely from one beam before exciting its closest neighbor, thereby avoiding overlap and confusion between the echoes; for example, it is best to "look away" for each successive firing. Finally, if an M-mode line (to be discussed in the next section) from one particular element is desired, this line can be updated several times within each scan time without waiting for an entire scan to be completed. This improves the time resolution of the M-mode measurement.

A principal disadvantage of the linear array pertains to the small size of each element forming the beam since each element must work alone. As discussed in Chapter 5 and shown in Figure 5.17b, the beam from a small element will diverge rapidly; if focused, it will not come to as small a size as can be obtained from a larger transducer. Thus, lateral resolution suffers from the small elements except at very short range.

Another way of stating this is to note that the transition distance $z_R = b^2/4\lambda$ from the near field to the far field is short for small elements, leading to beam divergence and poor focusing ability at depth. To overcome this with linear arrays, the elements could be made wider (i.e., increase the ℓ dimension) or perhaps several at a time could be excited. This extends the near-field region, giving less divergence and better focusing capability. The tradeoff to this action, however, is that neighboring scan lines in the linear array scheme can be no closer together than the spacing between the elements. In turn, the system's lateral resolution can be no better than the spacing between the scan lines since this is how individual scatterers are detected as being separate, namely, when they appear on separate lines. Thus, as the element width increases, the scan lines get further apart, ultimately limiting lateral resolution. Problem 6.9 is an exercise in finding the optimum element width of a typical unfocused linear array transducer.

To partially overcome the problem of poor lateral resolution, the elements may be grouped together when excited and when used as receivers (say, eight neighbors at a time) to increase the effective aperture. But when incrementing to the next scan line, only one outermost element is dropped from one side and one added to the other side of the group, giving an effective line spacing which approximates one element width. This scheme improves lateral resolution.

The linear array is popular, especially for abdominal and obstetrical scans. However, it is more electronically complex than a mechanical scanner due to the requirement that each element have its own channel of front-end amplification electronics and its own pulsing capability. Figure 6.15 shows a picture of a linear array and a typical image.

A potential problem arises when a long linear array is used to image the heart. As shown in Figure 6.14, in certain orientations the ribs will block some of the beams from the array, producing shadows (and multiple reflections) in those areas. The array head is by necessity quite long (perhaps 10 cm or more) to give a wide picture, and this makes its placement in the cardiac windows between ribs more difficult.

In the phased array, to be discussed next, the sector scan spreads out from the transducer in a fan shape, similar to that obtained with mechanical real-time B-scanners. Therefore, the overall transducer width can be smaller while still giving a wide view, making its placement in the cardiac windows easier.

6.4 REAL-TIME B-SCANNERS

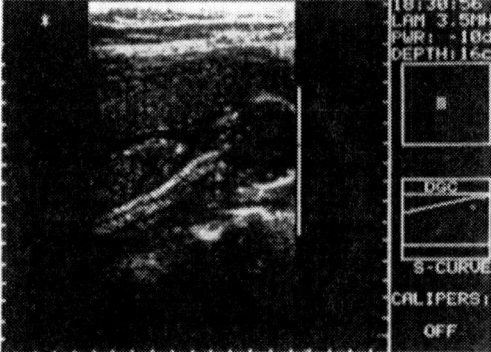

Figure 6.15 (a) Linear array transducer, especially useful for abdominal and obstetrical scans. (b) Typical fetal image produced by the linear array. Note the distinct image of the developing fetal spinal column. Photographs courtesy of Diasonics, Inc.

Phased Array

If the array elements are operated simultaneously, but with a controlled phase relationship between them, it is possible to electronically sweep the beam through all the sector angles and even to focus the beam in addition. Figure 6.16 shows how this is accomplished.

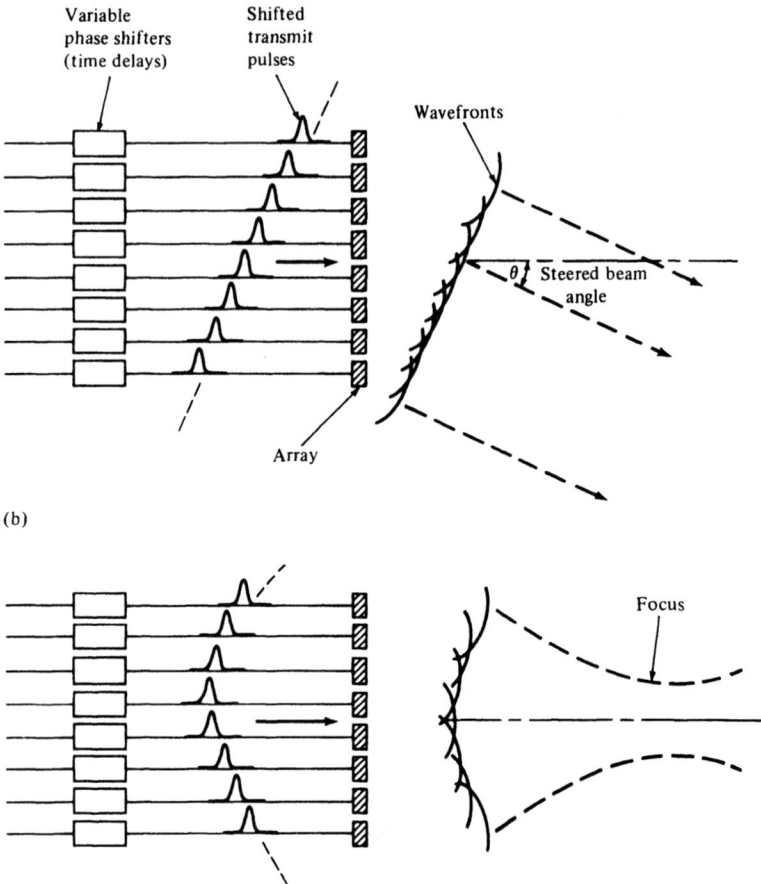

Figure 6.16 (a) By differentially delaying (phase shifting) the pulses to the element in a linear fashion across the phased array, the resulting transmitted beam can be electronically steered away from perpendicular. (b) By imposing a spherical time delay curvature across the array, the beam can be made to converge and focus at a desired depth. Combining (a) and (b) gives simultaneous beam sweeping and focusing. The principle works during the receive mode as well.

The first part of the figure portrays the situation in which the individual exciting pulses to the transducer elements are progressively shifted in time from each other by variable-time delay modules before they reach the array. This time shift is equivalent to a shift of phase, hence the name "phased array." In this case the time shifts follow a linear pattern across the array from one side to the other. The uppermost element will be excited first, and an approximately spherical acoustic wavefront (since the element is small) will begin propagating away from it. A very short time later, the next element will be excited and its wavefront will join the one from the previous element. The sequence will continue until the last element is excited and its wavefront is added to the total.

Now, by Huygen's principle the total wavefront is just the sum of its parts. Since the wavefront of the previous element will have progressed a certain distance before the next element is fired, the combined wavefront will be tilted off at an angle away from the perpendicular. At that angle the individual wavefronts from each element constructively interfere, forming a quasi-planar resultant wavefront. At other angles there will be various degrees of interference between the contributing waves, sometimes partial constructive interference (in the side lobe and grating lobe directions, for example) but mostly partial or complete destructive interference. This leaves a substantial beam that propagates only in the direction of tilt determined by the phase shift between elements.

The actual beam shape has already been derived in Chapter 5, and its angular pattern has been shown in Figure 5.24. The only modification needed here is to tilt the solid pattern by an angle given by the amount of phase shift. (Note, however, that the dashed envelope shown in Figure 5.24 is not tilted, since it is due to the radiation pattern of the individual elements, which, of course, are not physically rotated. In fact, this envelope is just the shape of the individual wavefronts radiated by each element.) Since all the elements of this array are used coherently, the pattern will have grating lobes as well as side lobes, and, as discussed earlier, the presence of these grating lobes will cause some off-axis sensitivity and artifacts in the return. We shall return to this question shortly.

Although Figure 6.16 deals with the transmitted beam explicitly, the concept of beam steering by controlling the phases of each element applies just as well when the array is used as a receiver. Reciprocity assures that the beam pattern of the receiving mode can be electronically steered using the same linear phase shift as for the transmitted beam. The only difference is a practical one: In the transmit mode, only constant-amplitude pulses of voltage are traveling to the transducer elements, so phase shifting can be accomplished by rather simple digital time delay circuits which are inexpensive and easily controlled. In the receiving mode, however, the phase shifting must be done for small *analog* echo voltages traveling the other way (after usually some preamplification). Therefore, the receiver phase shift modules must be able to handle analog voltage waveforms, preserving phase and amplitude over a wide dynamic range while performing the variable phase shifting without introducing appreciable noise. This formidable task requires specialized circuitry; and since one phase shift module is needed for each channel of input, this adds considerably to the cost and complexity of phased array instruments. In the receive mode, the shifted signals are summed together after phase shifting and some initial signal conditioning to produce a single output.

There are basically two ways to accomplish the analog phase shifting in the receiver channels. One is to use true analog phase shifters, such as tapped lumped-element transmission lines or integrated circuit sampled-capacitor serial analog memory modules (SAM) or charged coupled devices (CCD).

The other technique, which gives more flexibility, is to first digitize the analog signal using a very fast analog-to-digital converter (A/D), then to phase shift the signals during their recombination by appropriate selection of the starting addresses of the temporarily stored digital signals. The A/D converter must come before the demodulation stage and must have high speed since the phase of each of the incoming signals must be preserved for proper beam steering. From sampling theory, the A/D converter should sample at a rate at least twice as high as the highest frequency to be preserved in the signal; for a 3-MHz echo, this requires a sampling rate of more than 6 MHz. In addition, the A/D converter needs a large dynamic range (60 dB) unless some signal compression precedes it. Since each channel requires its own A/D converter, the system cost quickly multiplies.

Focusing the Phased Array

Returning now to Figure 6.16, the lower portion of the figure shows how focusing of the transmitted beam can be achieved. By imposing a spherical time delay shape on the pulses exciting the elements, the resultant outgoing wavefront is made concave, causing the beam to converge to a focused spot at a distance determined by the severity of the phase curvature. This focusing ability significantly improves the lateral resolution of the phased array beam compared to a linear array beam, since the entire array width L determines the focused spot size, not just a single element width b.

Since only one focal length is possible for each transmitted scan line, it is usually positioned near the mid-depth of the region of interest. (Note, however, that the flexibility of electronic focusing allows this position to be different for neighboring lines, giving staggered transmit focal lengths among the lines for better resolution throughout the depth; some instruments take advantage of this capability.) The minimum focused spot size (tight focusing) is given by Equation (5.52), with b replaced by the array width L; and due to the relatively large value of L, the focused spot can be small. This gives excellent resolution at the position of the focus, but at other positions along the scan line the resolution degrades as the beam expands. The smaller the focused spot size, the shorter is the range (depth of focus) over which the beam stays small; Figure 5.22 showed this effect.

To strike a balance between resolution at the position of focus and at other locations along the scan line, the transmit focus is sometimes purposely made nonideal (weak focusing). This can be done, for example, by giving the transmitted wavefront a nonspherical concave shape, leading to an extended line of focus that is wider than the diffraction limit in the transverse direction but that extends for a much greater distance along the scan line; this is shown in Figure 6.17a. The curvature of the transmitted wavefront is easily adapted to the desired aspherical shape by tailoring the individual channel time delays.

The situation is even more flexible during the receive mode of each scan line. Whereas only one focus is possible when transmitting the outgoing pulse, several focal lengths can be employed sequentially during the time of receiving each scan line. In fact, by continually updating the focal time delays to the analog phase shifting modules, the focused point of the receiver can *follow* the transmitted pulse as it propagates into the tissue, always

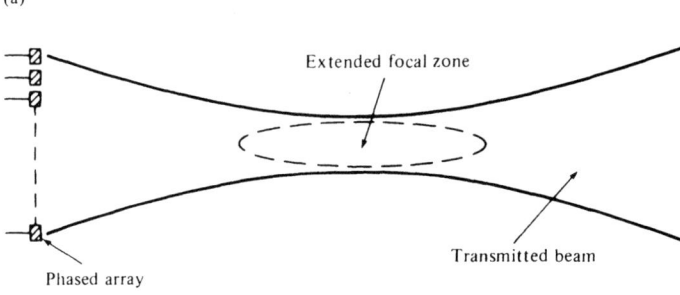

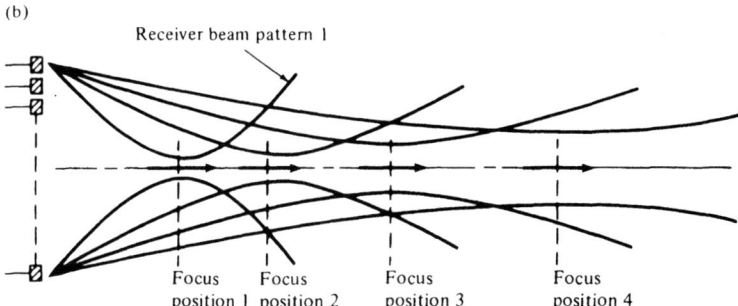

Figure 6.17 (a) Using an aspherical time delay curve, the transmitted beam can be focused to an extended focal zone, increasing the depth of field. (b) During the receive mode, the focal position can be stepped progressively outward, following the echoes from the propagating transmit pulse.

keeping the receiver in focus with the ever-increasing depth at which the echoes are being produced. This capability of electronically sweeping the receiver's focus within a scan line is called *dynamic focusing* and is unique to the phased array system.

In practice, the point of receiver focus is not continuously moving. Because it takes a finite amount of time to send and load the proper time delay information to each of the analog channels, the focal length is updated only in discrete steps. It is usually adequate to have a restricted number of focal zones, spaced so that they meet or overlap at the ends of their respective depth of focus. Within each focal zone, the focused spot is made as small as possible (because dynamic focusing alleviates the worry of a stationary, limited depth of focus) by using spherical phase curvature across the array. Since the depth of focus is proportional to the square of focal length (see Equation (5.55)), the distance between focal zones can get farther apart as the focus moves away from the array. Figure 6.17b shows four such successive zones.

Phased-Array Time Delay Equations

The equation for the time delays needed in each of the channels to achieve steering and focusing is derived by reference to Figure 6.18. Let P be the

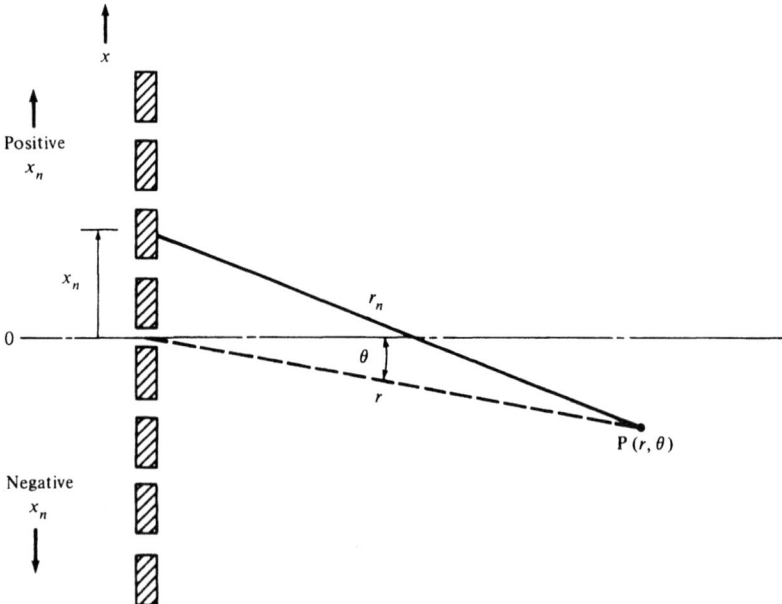

Figure 6.18 The geometry for deriving the element time delays needed for focusing and steering the beam to point P.

desired point of steering and focusing; r is the distance from the center of the array to P, and r_n is the distance from the nth element to P. Since r_n is dependent upon element position, appropriate time delays will have to be incorporated in each channel to ensure that the pulses from all elements will arrive at P at the same time for constructive interference (transmit mode) or, equivalently, that an echo from P will be in phase in all channels when summed (receive mode). By trigonometry,

$$r_n = (r^2 + x_n^2 - 2rx_n \sin \theta)^{1/2} \tag{6.5}$$

and the difference between the reference distance r and r_n is

$$\Delta r_n = r - r_n = r - (r^2 + x_n^2 - 2rx_n \sin \theta)^{1/2} \tag{6.6}$$

By the Fresnel approximation, when $r \gg x_n$, Equation (6.6) may be written as

$$\Delta r_n \approx x_n \sin \theta - \frac{x_n^2}{2r} \tag{6.7}$$

Since time of propagation is related to pathlength by

$$\text{time} = \frac{\text{distance}}{\text{velocity}}$$

the time delay for the nth channel is

$$\boxed{\Delta t_n = \frac{\Delta r_n}{c} = \frac{x_n \sin \theta}{c} - \frac{x_n^2}{2cr}} \tag{6.8}$$

This equation gives the time delay which must be imposed for each channel n to obtain constructive interference at the point $P(r, \theta)$. Note that for some circumstances, Equation (6.8) requires a negative time delay. This will happen when the distance from element n to the point P is greater than the reference distance r. Since negative time delays are not physically realizable, a constant-bias time delay is merely added to all channels such that the required delays will be positive for all focal point positions.

It is significant that Equation (6.8) separates into two distinct terms under the Fresnel approximation. The first term,

$$\Delta t_n^s = \frac{x_n \sin \theta}{c} \tag{6.9}$$

represents the amount of delay needed to *steer* the beam's axis to angle θ; it is independent of r and is linear in terms of x_n. The second term,

$$\Delta t_n^f = \frac{-x_n^2}{2cr} \tag{6.10}$$

gives the delay necessary to *focus* the beam to radius r; this term is independent of θ and is quadratic in x_n, leading to a "spherical" time delay curve across the array. Thus, it can be seen that the two functions, steering and focusing, can be accomplished simultaneously and independently under this approximation by simply adding their individual contributions:

$$\Delta t_n = \Delta t_n^s + \Delta t_n^f \tag{6.11}$$

This simplifies the design and dynamic controlling of the machine. The Fresnel approximation which leads to the independence of the terms is accurate for beams steered at small angles but less so for beams at the edges of wide-angle sectors.

Characteristics and Advantages of the Phased Array

After summing the channels, the processed echoes from the phased array are displayed in normal B-mode fashion. Due to the nature of the beam sweeping, the sector appears pie-shaped, with many scan lines radiating from the apparent center of the array as shown in Figure 6.19.

To be effective as a real-time imager for echocardiography, each complete sector should be obtained within a short time with a minimum sector repetition rate of between 30 and 60 frames per second. This sets a limit on how deep the sector can view for a given number of lines or,

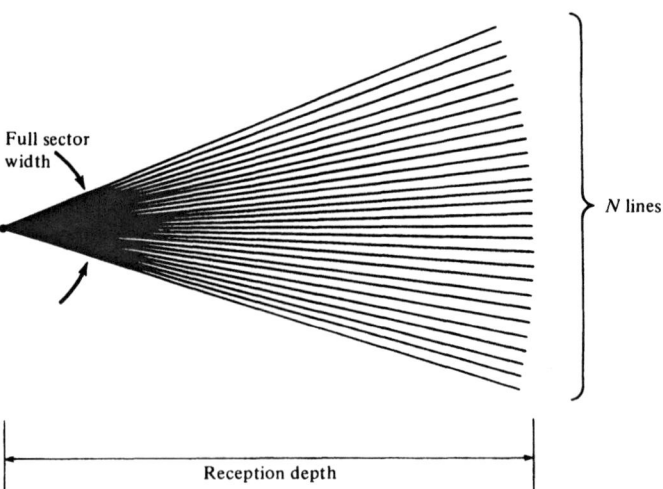

Figure 6.19 The pie-shaped display of a phased array is characteristic of B-mode sector scanners. Many phased array instruments allow a selectable tradeoff between depth, number of lines, and line density (through selection of the angular sector width).

6.4 REAL-TIME B-SCANNERS

conversely, how many lines can be obtained for a given depth. The optimum tradeoff between depth of view and number of lines will depend upon the particular application, so many machines allow a selection of depth choices and corresponding line numbers.

In turn, the line density (how close together the lines are), which may set the ultimate limit on lateral resolution, depends upon the total number of lines within the sector and the full angular width of the sweep. Therefore, these machines will also allow a selection of various angular sweep widths (typically 30°, 60° and 90°). The large angular views are used for initial observation and orientation, while the narrower views will give a higher line density for a detailed structure analysis.

Due to its unique characteristics, the phase array is well suited for cardiac imaging. Like the linear array, it possesses all of the flexibility of electronic beam sweeping, such as speed of steering, the ability to intermix an M-mode line with the B-scan, nonregular scanning sequences for reducing late echo sensitivity and the windshield wiper effect, and a transducer head that is free from mechanical components.

The phased array has additional advantages over the linear array for imaging the heart. Since the entire array width is used coherently, the beam size in a phased array is smaller than that in the linear array and lateral resolution is improved, especially when dynamic focusing during the receive mode is employed. The sector lines fan out from the center of the array, allowing a smaller overall transducer width to get a large area view. This smaller overall transducer size (typically 2 to 4 cm) will fit more readily into the limited ultrasonic access to the heart between the rib structure and lungs, allowing better placement and acoustical contact while maneuvering to see various cardiac structures. Figure 6.20 shows a modern phased-array instrument, along with a display typical of those obtained by phased-array echocardiography.

There are two main disadvantages of the phased array. Because every channel needs a controllable analog phase shifter and separate preamplifier, the system complexity and cost are relatively high. Also, the beam pattern from the array contains grating lobes giving unwanted off-axis sensitivity. The grating lobe amplitude is reduced somewhat in the transmitted beam when short pulses are used to excite the elements (see Section 5.6), but they still are an undesirable artifact. The angle of the grating lobes can be moved further away from the main beam by decreasing the element spacing or increasing the wavelength, but other specifications of the instrument will then suffer (see Problem 6.11).

It should be mentioned that the phased-array principle will also work with concentric ring (bull's-eye) transducers. In this case beam steering is not possible, but focusing during transmit and receive can be accomplished. Due to the limited number of rings, this system will be correspondingly less complex. Beam steering, if needed, may be done with

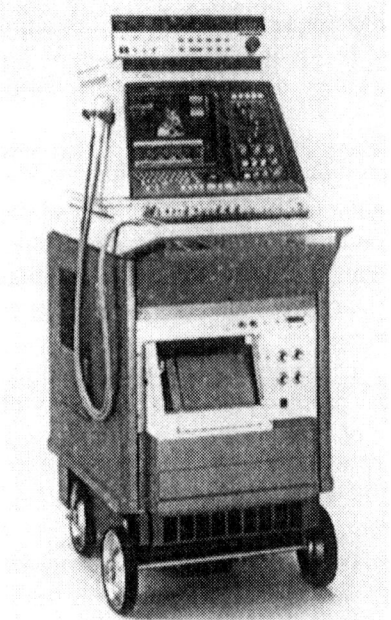

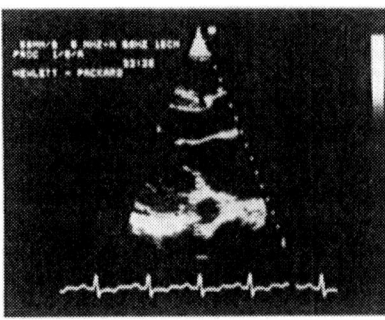

Figure 6.20 (a) A phased array instrument, especially useful for real-time cardiac imaging. (b) A B-sector view of the heart, produced by the phased array machine shown in part a. Photographs courtesy of Hewlett-Packard Company.

mechanical means. Also, phased array focusing may be employed as part of a hybrid linear array transducer.

6.5 M-MODE

This configuration is used for analyzing, both qualitatively and quantitatively, the motion of body structures such as heart valves. The "M" stands for "Motion," and this mode is sometimes referred to as the TM-mode, for "Time Motion." Figure 6.21 is a diagram of the essential elements of

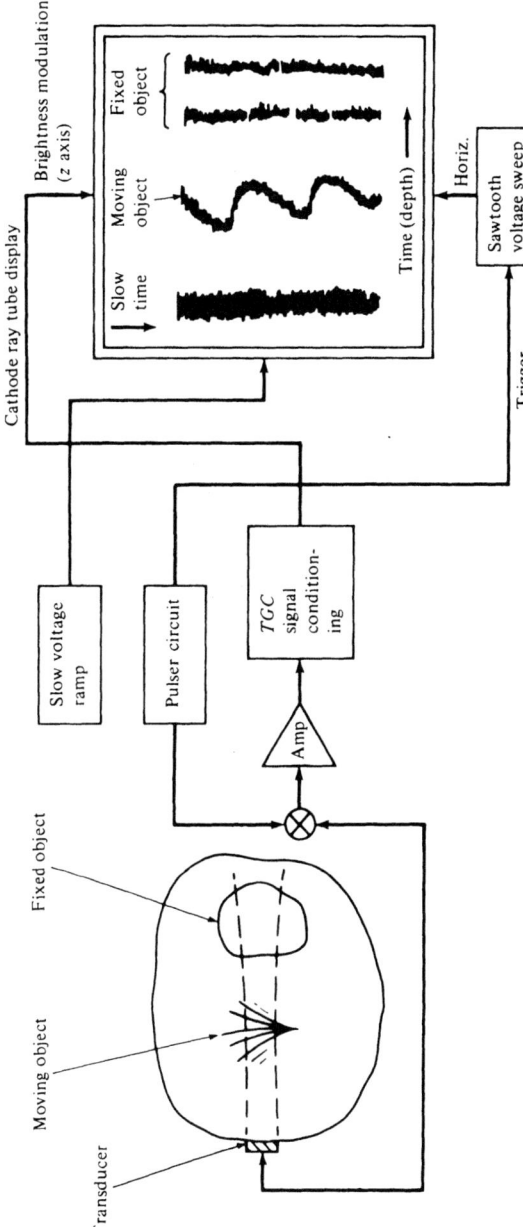

Figure 6.21 An M-mode instrument. The transducer beam is kept stationary while the echoes from a moving reflector (such as a heart valve leaflet) are received at varying times sequentially displayed as changing positions of the brightness-modulated dot corresponding to the echo. The vertical deflection of the display is swept at a slow rate so that perhaps two or three cardiac cycles are contained in a single display. Net displacements and velocities may be obtained by measuring the display.

an M-mode configuration. It is a hybrid mode with some characteristics of A-mode and some B-mode. As in the B-mode the brightness of the display line is modulated according to the amplitude of the received echoes. It is similar to A-mode, however, in that echoes are collected only in one dimension, along the steadily held direction of the beam. These signals are presented on the horizontal axis of the display.

The vertical deflection of the display is controlled by a slowly varying linear voltage ramp such that successive lines are written in progressive order down the display, as on a printed page. Any movement of an object along the beam's path will show up as a horizontal displacement of the object's echo recorded by the successive lines and appears as a waviness in position of the corresponding echo dots.

The horizontal sweep time is identical to that found in the A-mode and B-mode, that is, 13 μs for each centimeter of depth traveled. The vertical sweep is much slower, about 2 to 3 s to cover the entire screen so that several heart cycles may be displayed. Since the display is calibrated in terms of depth on the horizontal axis, the net spatial displacement of the moving object can be measured directly. This is a valuable aid to determining the extent of net movement in heart valve leaflets. Also, since the vertical axis is given in units of time, the velocity of the object may also be measured quantitatively from the display, in units of mm/s (see Problem 6.14).

Permanent recording may be done photographically or on video tape, as before, or may be obtained by optically exposing self-developing strip-chart paper via a flat optical fiber cable from the display to the recorder. Many of the B-scan linear-array and phased-array instruments intended for cardiac imaging have the feature of allowing a simultaneous M-mode image to be obtained along one of the lines of the sector selected by the operator. The M-mode line is updated at a rate faster than the full sector is swept to improve the velocity resolution of the M-mode. Figure 6.22 shows an example of an M-mode record.

6.6 C-MODE

This mode differs from all of the previously described modes in that it does not rely upon the detection of reflected echoes from interfaces. Rather, C-mode generally refers to *through-transmission* imaging in which the ultrasound pulse is transmitted from one side of the body through to receiving transducers on the opposite side.

Figure 6.23 shows the basic components needed for the simplest level of C-mode scanning. A single transducer acts as transmitter propagating

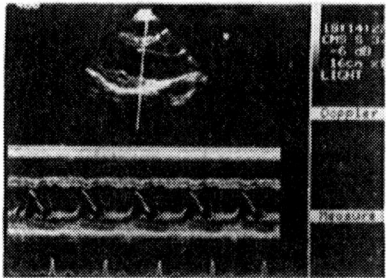

Figure 6.22 (a) An M-mode image shown near the bottom of the picture (below a B-mode mechanical sector scan). The selectable M-mode line is highlighted on the B-mode image. An EKG tracing appears at the very bottom of this picture. (Note: the orientation of this M-mode record is rotated 90° from that shown in Figure 6.21.) (b) A view of the machine which produced the image of part a. It is also capable of linear array and Doppler imaging. Photographs courtesy of Diasonics, Inc.

a pulse through the region being imaged. A separate receiving transducer detects the delayed and attenuated pulse. Transverse spatial resolution at any position is determined by the product of the overlapping beam patterns of the two transducers. Scanning is usually accomplished by translating the two transducers together as a pair affixed to a rigid yoke spanning the body. Since the scanning motion is in a plane perpendicular to the beam, a two-dimensional image can be obtained similar to the anterior–posterior images produced by conventional X-ray films (see Figure 1.1).

Two tissue characteristics can be derived from the information received in C-scans. First, by comparing the amplitude of the received pulse to the amplitude of the transmitted pulse, a measure of the total attentuation along the path may be made. This attenuation includes not only absorption by the intervening tissues, but also reflection losses at the interfaces encountered and scattering along the path.

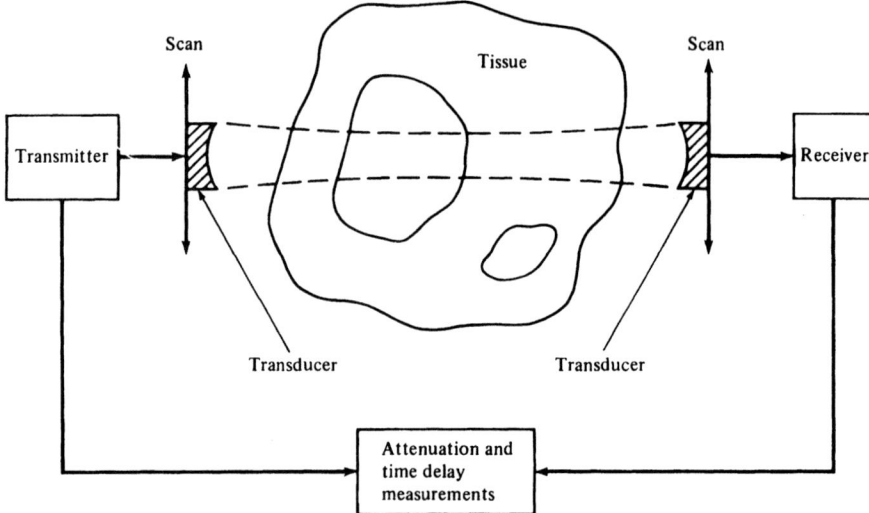

Figure 6.23 Basic configuration for C-scan (or through-transmission scanning). Both attenuation and velocity data may be obtained by this method. If the scanning motion includes both translation and rotation over a large range, enough information is gathered to allow computerized reconstruction of the two-dimensional tomographic image, analogous to X-ray CT scanning. A water bath or water coupling bags are used to facilitate scanning. In advanced designs the scanning motion may be supplanted by large arrays of transducer elements.

Second, by using a carefully stabilized path length, a comparison of the time between pulse transmission and reception ("time of flight") will give the data needed to calculate phase velocity of the tissue in the path; for a fixed distance the time delay is inversely proportional to velocity. The phase velocity properties of a material may be normalized to the velocity of sound in water by defining its *acoustical index of refraction* to be

$$n = \frac{c_{\text{water}}}{c} = \frac{1.5 \times 10^5 \text{ cm/s}}{c} \qquad (6.12)$$

Thus, the relative time delay through the material will be proportional to its refractive index. Unlike the optical index of refraction (which is normalized to the speed of light in a vacuum and therefore can never be less than unity), the ultrasound index of refraction can take on values above and below 1.0.

C-Scans will therefore give information regarding both the attenuation and index of refraction along the path. However, since the ultrasound pulse must travel the entire distance through the tissue in C-scans, regions of complex anatomy are generally difficult to image because of the multiple reflection losses and refracted angle changes undergone by the pulse. Also, total attenuation is high and received signal-to-noise ratios tend to be low.

The best success with C-scans has therefore been in imaging relatively homogeneous anatomy such as the female breast from the lateral directions. In the breast, lesions or cysts often show up as areas of increased attenuation and/or modified index of refraction. To facilitate the mechanical scanning motion required in the C-mode while maintaining acoustic coupling, the breast and transducers are both placed in a warmed water bath, or coupling is provided by means of sealed water bags.

We now investigate some advanced imaging techniques that fall into the broad category of C-scanning.

6.6.1 Acoustic Holography and Synthetic Aperture

The term holography means "whole picture." When applied to imaging by waves (either optical or acoustical) reflected or transmitted by an object, it indicates that not only the amplitude but also the *phase* of the waves is recorded; this preserves the entire information possible regarding the interaction of the waves with the object. Since measurement of phase is equivalent to a measurement of time delay, holography and determination of refractive index are related techniques.

In optics, the frequency of the illumination is so high (approximately 10^{14} to 10^{15} Hz) that the only practical way to record the phase of an object's waves is to combine them with a known reference wave from a narrowband source such as a laser, and to rely upon constructive and destructive interference to encode the phase information as interference fringes on a piece of film. The striking three-dimensional reconstructions from these optical holograms are evidence of the fact that recording phase along with amplitude does indeed give a whole picture of the object.

Spurred by the success of optical holography and with narrowband sources of acoustical energy readily available in the form of cw ultrasonic transducers, researchers pushed the development of acoustical holography in the late 1960s and early 1970s. Following the lead of the optical arrangement, the early acoustical holography studies combined the waves from the object, usually in the through-transmission mode, with a reference wave obtained either as a portion of the original source wavefront or generated by a separate transducer locked in phase with the source transducer. The recording of the resultant interference pattern was achieved by a variety of techniques, some of them quite involved, such as optically scanning the pattern of interference ripples that occur on the surface of the water tank with a laser beam and detecting the deflections of the beam. Figure 6.24a shows one general configuration for holography.

(a)

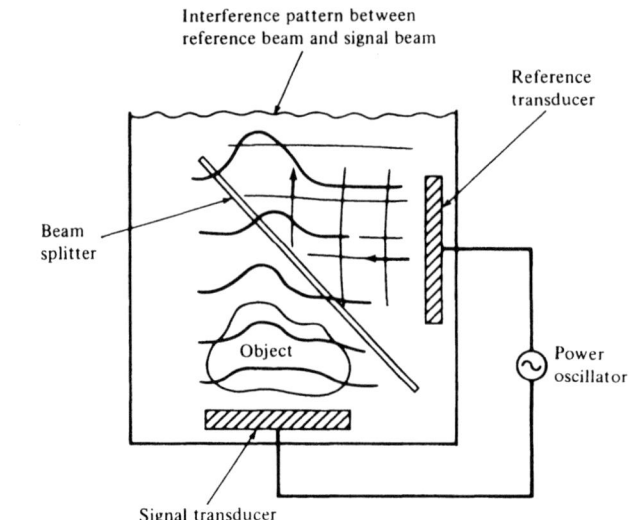

(b)

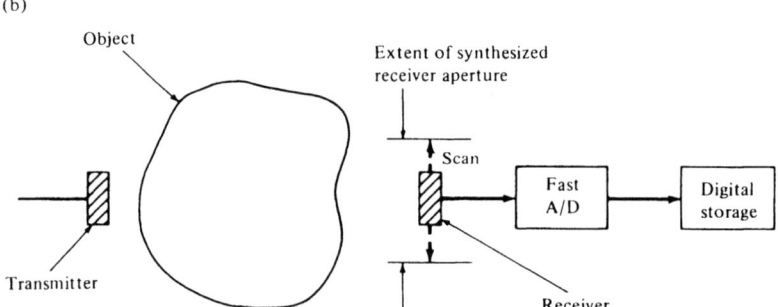

Figure 6.24 (a) In *acoustical holography* the amplitude and phase of the waves from an object are encoded in the interference pattern produced by combining the object (signal) waves with a separate reference wave. (b) The *synthetic aperture* technique stores the received signals (amplitude and phase) from several receiver positions. By proper recombination this will be equivalent to an effectively larger receiver aperture for improved lateral resolution.

Holography with sound waves is quite successful for some objects, such as for relatively homogeneous plastic or metal fabricated parts, where it is used as a means of nondestructive testing for otherwise hidden flaws and cracks. However, when applied to biological objects, only the simplest regions (such as limbs) give acceptable pictures. When through-transmission acoustical holography was initially clinically tested for abdominal imaging, there were so many multiple reflections and phase perturbations in the complex tissue structures of the abdomen that the resultant pictures were difficult to interpret in clear anatomical terms.

Studies are still continuing in this area, but recent developments have emphasized another way of recording the phase information. The frequency of ultrasound (10^6 to 10^7 Hz) is much lower than that of optical waves, and there exists the possibility of directly detecting and recording the phase of the object's waves without using interference patterns. The voltage waveform from a broadband transducer can follow the phase of the received waves; and with recent advances in the speed of A/D converters, this voltage waveform, including phase, can be digitized at a rate of several MHz and stored in digital memory. If this is done for a series of scanned transducer positions the stored signals can be combined after reception in any phase and amplitude fashion desired.

For example, summing the received waveforms from a wide scanning area gives a result equivalent to that produced by a single large receiver with dimensions equal to the overall extent of the scanned area, not the smaller sampling transducer. This large *synthetic aperture* gives significantly improved lateral resolution. In addition, by imposing phase differences between the stored samples before summing, a focusing lens reminiscent of the phased array can be synthesized. Since these data manipulations are done under computer control after reception, there is the possibility for complex processing of the signals. Figure 6.24b shows a sketch of the general configuration for a synthetic aperture.

This technique requires increased time for performing the scan and needs a fair amount of memory and a fast computer, but some day clinical instruments may incorporate synthetic processing of their received signals.

6.6.2 Ultrasonic Computerized Tomography

The tissue quantities measured at any position of the transducers in basic C-scan are values integrated along the entire path through the body. To achieve any axial resolution regarding individual contributors along the path, additional steps must be taken. One way to provide axial resolution is to use a focused receiver (and perhaps transmitter) whose focal position is moved along the path. With sharp focusing, the region within the focal zone is the largest contributor to the received signal, and therefore the sensitive region may be scanned in the axial direction.

Another way to provide axial resolution is by using computerized reconstruction, analogous to X-ray CT scan techniques. Here the scanning motion of the transducer pair is taken laterally through all possible paths as well as rotated around the body in that plane. By storing the integrated data from all these possible paths, the computer can then unfold the information from each small subregion ("pixel") within the scanned plane by using one of various reconstruction algorithms. The resulting two-dimensional image is a tomogram analogous to the view from a CT scan. Much research attention is now being given to appropriate algorithms for such ultrasonic computerized tomographic scanning.

One difficulty encountered with ultrasound not found in X-ray computerized tomography is the refractive bending of the transmitted beam as it passes through regions of differing phase velocity. It may be possible to compensate for this effect by iteratively constructing phase velocity maps of the region with time delay data provided by the scan, since from Snell's law the amount of beam refraction is related to phase velocity changes.

6.7 OTHER MEDICAL USES OF ULTRASOUND

Although imaging and Doppler velocity measurements are the mainstay of ultrasound applications in medicine, a few other techniques employ the use of ultrasound to analyze, detect, or modify tissues in the body. These relatively new directions in the application of sound waves to medicine are not yet nearly as advanced as the imaging and Doppler fields, but research and clinical trials are continuing to test their usefulness. We mention three examples: ultrasound surgery, hyperthermia production, and the acoustical microscope.

6.7.1 Surgery with Ultrasound

The mechanical action of molecular vibration that accompanies the passage of ultrasound can in some circumstances be used to disrupt or dislodge unwanted tissue or particles. For effectiveness, the power density must be high, so to avoid damage to nearby normal tissue the ultrasound is usually applied by a localized applicator or with considerable focusing.

For example, cataracts in the lens of the eye may be fragmented and removed by an ultrasonic probe which consists of a small vibrating needle connected to a transducer in a handpiece. The frequency of vibration is approximately 40 kHz. The probe is inserted through small incisions in the cornea and the tip positioned at the site of the cataract. Vibrations of the tip disrupt the structure of the cataract, and the pieces are aspirated through the lumen of the needle by suction. Irrigation may also be used to clear the products of the fragmentation. The advantages of this technique over conventional surgery are that smaller incisions are possible and the anterior chamber and cornea of the eye are less disturbed.

A similar approach is finding success in breaking up stones in the kidney and upper urinary tract. When stones in these areas become too large to be naturally (and painfully) passed through the ureter, they must be surgically removed. One alternative to open surgery is percutaneous access to the stones by inserting a probe through the skin to the site of the calculi in the kidney's collecting system. A fiberoptic endoscope (nephroscope) is then inserted in the probe to allow viewing the stones while small forceps and baskets are manipulated to pull out the smaller stones through a working lumen in the nephroscope.

If a calculus is found that is too large to be withdrawn through the probe's channels, it must be broken up (a process called *lithotripsy*). This may be done with ultrasound applied by a long hollow needle attached to an external transducer. The ultrasonic needle is passed down the probe and contacts the rigid matrix of the stone. Vibrations at a frequency of approximately 25 kHz gradually disintegrate the stone, and the fragments are removed by suction through the center of the needle. The percutaneous ultrasonic procedure can circumvent the need for open surgery in the case of large stones, and this means fewer days of hospital care and quicker recovery. Complications are also reduced.

Another ultrasonic way for breaking up stones is currently employed clinically that does not use invasive probes or needles to carry the ultrasound. Rather, the sound waves are propagated into the kidney through the overlying tissue from a high-energy shock wave produced external to body. The shock wave is produced in water by a high-voltage spark placed at one focal point of a large ellipsoidal reflector. At the other focal point, the patient and stone are precisely positioned with X-ray views. The highly concentrated sound waves disrupt the rigid structure of the stone apparently without causing undue damage to the soft tissues surrounding it. From 500 to 800 shock waves may be needed to adequately fragment the stone, which then is naturally excreted with the urine.

The advantage of this approach is that no invasive probes are needed, and surgery is avoided. One disadvantage is the high cost of the equipment and occasional discomfort for the patient.

6.7.2 Heat Production and Hyperthermia

The absorption of ultrasound will transfer energy from the propagating wave into heat in the absorbing tissue. This is purposely kept low for diagnostic imaging machines, but in some circumstances the heating ability is used to advantage. One application is in ultrasonic diathermy for treating sore muscles and joints, especially in athletes.

The amount of power deposited in the tissue may be determined from the equation describing the propagating wave intensity. From Chapter 4,

$$I(z) = I_0 e^{-2\alpha z} \tag{6.13}$$

where α is the pressure attenuation coefficient. In an incremental distance dz, the power lost in the wave is found by differentiation:

$$\frac{dI(z)}{dz} = -2\alpha I_0 e^{-2\alpha z} \tag{6.14}$$

which gives the rate of power loss per unit volume at a distance z from an initial intensity of I_0.

The attenuation coefficient α accounts for both scattering and absorption, but in most homogeneous tissues the absorption greatly predominates. Therefore, Equation (6.14) gives a good estimate of the rate of heat generation. Note that the attenuation coefficient α, which is approximately proportional to frequency for many soft tissues, enters the equation in two places. For high-power deposition, the attenuation should be large, of course, but that also decreases the amount of power left in the beam after it has penetrated through intervening tissue, as the negative exponential term in Equation (6.14) dictates. Thus, when it is desired that the power be deposited at deep sites, there is an optimum frequency which maximizes the amount of deep heating (see Problem 6.13). Typical diathermy frequencies range from 1 to 3 MHz at intensities near 1 W/cm^2.

A promising new research area for cancer therapy is hyperthermia. In this technique, the body region containing the cancerous tumor is raised in temperature above normal by several degrees (to 43°C or more, if possible). This elevation in temperature lasts for 30 to 60 minutes per treatment and is usually preceded or followed by conventional treatment for the cancer such as radiation therapy. Initial clinical results appear encouraging for some tumor types.

The heating of the cancerous region in many cases is accomplished by electromagnetic means, such as by radio-frequency inductive or capacitive applicators, or by microwave radiation. In some areas of the body, though, the much shorter wavelength of ultrasound will give a more precisely defined focused spot size and better control of the power deposition pattern than with electromagnetic waves. This is especially advantageous for small tumors located in relatively homogeneous tissue regions such as the breast, brain, and perhaps the soft tissue areas of the head and neck.

Therefore, some hyperthermia treatments may be best executed with ultrasonic heating. The areas of the body where ultrasound cannot be successfully employed are those where bone or air regions block its path.

6.7.3 Ultrasonic Microscope

As discussed in Chapter 5, the ultimate resolution of imaging instruments, including microscopes, is set by the wavelength of the illumination used. Good optical microscopes can therefore resolve features of slightly less than 1.0 µm; they are common and easy-to-use devices. For better resolution, electron microscopy can be employed. With high-energy electrons, the resolution can approach several nanometers (10^{-9} m).

A microscope using acoustic waves has also been developed. Because the viewed objects are generally very small samples, their short absorbing path length allows the use of ultrahigh frequencies for good resolution. Currently, frequencies as high as 1500 MHz are employed; the wavelength and approximate resolution in water at this frequency is 1.0 µm.

Since the resolution of the ultrasonic microscope is not significantly better than that of the optical microscope, what place does it have in imaging? First, it uses the natural aqueous environment surrounding the sample to facilitate acoustic coupling, and since the sound waves do not appreciably disturb the sample, it can image living substances such as cells to follow their movement and other biological properties.

More importantly, the acoustic microscope images different characteristics of the sample than either the light or electron microscopes. Used in the reflection mode, it is sensitive to patterns of acoustic phase and impedance interfaces within the object; in the transmission mode, it gives pictures of acoustic attenuation and refraction. For some substances, such as soft tissues, the structural variations within the sample may show up more dramatically in its acoustical properties than in an optical view. In short, the ultrasound microscope gives another view of the world.

Applications also include imaging the details of integrated circuits and metallurgical grain boundaries. As experience with the technique grows, applications will broaden also.

PROBLEMS

6.1. (a) For the anatomical configuration shown below, sketch the arrival times of all echoes as they would be seen on an A-mode display. Do not calculate echo strengths, merely calculate times of arrival referenced to the initial transmit time, out to a maximum of 200 µs. Include multiple reflections and use the precise phase velocity of the materials shown.

(b) What percentage error is introduced in the calculated arrival time of the first echo if the phase velocity is assumed to be the average calibration value ($c = 1.54 \times 10^5$ cm/s) rather than fat?

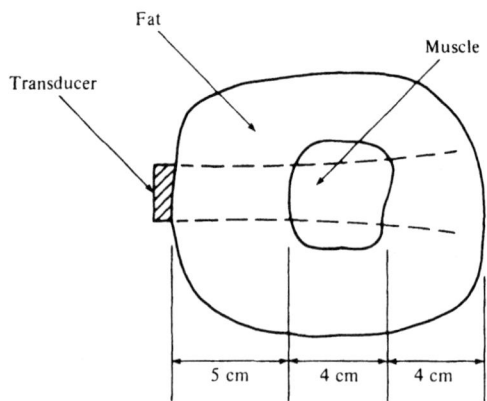

6.2. The A-mode display for a certain body region is shown below. Sketch two different anatomical configurations for the region which are consistent with the display obtained, giving distance values. Assume $c = 1.54 \times 10^5$ cm/s.

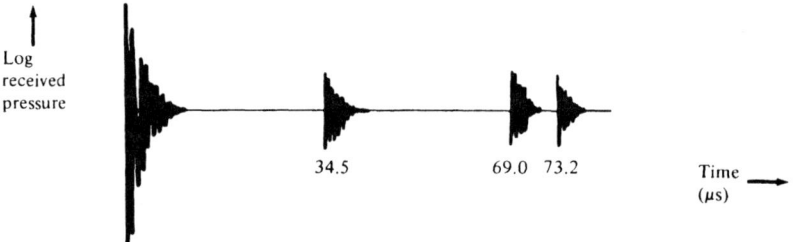

6.3. (a) Derive Equation (6.4) assuming that the tissue loss may be approximated as 1 dB/cm MHz.

(b) Calculate an appropriate rate of *TGC* increase using Equation (6.4) for a transducer whose fundamental frequency is 2.25 MHz, and determine the overall *TGC* increase which would be needed for a depth of travel to a boundary 15 cm deep. Can this range be reasonably achieved by a standard amplifier?

6.4. (a) Show that Figure 6.5b gives the correct shape for the frequency spectrum $|F(f)|$ for the rectangular pulse in part a of that figure. Use the Fourier transform relationship to find the frequency spectrum. The cosine portion of the Fourier spectrum is given by

$$F_1(f) = \int_{-\infty}^{\infty} p(t) \cos(2\pi ft) dt$$

and the Fourier sine portion is given by

$$F_2(f) = \int_{-\infty}^{\infty} p(t) \sin(2\pi ft) dt$$

The total Fourier magnitude is found from

$$|F(f)| = \sqrt{F_1^2(f) + F_2^2(f)}$$

(*Hint:* To simplify the calculation, let the pulse envelope of the cosine pressure wave start at time $-t'/2$ and end at $+t'/2$. This makes the entire waveform an even symmetric function of time, and one of the two Fourier portions will equal zero. Also, only consider the frequency spectrum for positive frequencies $f > 0$.)

(b) Estimate the width of the spectrum by finding the expression for half the frequency spacing between the two zeros on either side of the main peak. This is a measure of the spectral width Δf.

6.5. (a) Redo Problem 6.4a for the case of a single pressure pulse whose wave form is a sinusoidal of fundamental frequency f_1 modulated by an exponentially decaying envelope, as shown in Figure 5.7. The pulse can be represented by

$$p(t) = \cos(2\pi f_1 t)e^{-t/\tau} \quad t \geq 0$$
$$p(t) = 0 \quad t < 0$$

(b) For a 2-MHz transducer with a Q of 5, make a rough plot of the positive frequency spectrum of this pulse.

6.6. Design the θ-beam steering module in Figure 6.10 for a noncompound hand-scanned B-mode instrument. Assume that the angle of tilt of the transducer head is sensed by a rotary potentiometer such that a voltage V_a linearly proportional to this angle (positive voltage on one side of the central direction, negative voltage on the other) is fed to your module. At a tilt angle of $+20°$, $V_a = +2.0$ V. In particular:

(a) Give the mathematical expressions that describe the desired horizontal sweep and vertical sweep voltage generated by your module to properly deflect the display's electron beam in response to V_a. Assume that the sensitivity of the display sweep circuitry (both horizontal and vertical) is 1.0 cm of deflection for each 0.5 V of input. Make the display full scale (i.e., 1:1 correspondence with tissue dimensions). Be sure to include the proper time dependence in your equations; assume that each line on the display is triggered at the instant of the initial bang ($t = 0$). With zero input voltages, the display is centered at the middle of the left margin of the screen. [*Hint:* Use vector decomposition to get the two components (vertical and horizontal) of a line at an angle θ.]

(b) Sketch a rough block diagram of an electronics module that will generate the expressions derived in part a, with the voltage V_a as an input.

6.7. A simplified model of a compound B-scan is shown below in which a single point scatterer is imaged by the transducer from two different directions (position A and position B).

How far apart will the point appear on the display screen as viewed from the two directions? That is, how much misregistration will occur between the two images of the point as seen in the overlapping sectors? Assume that

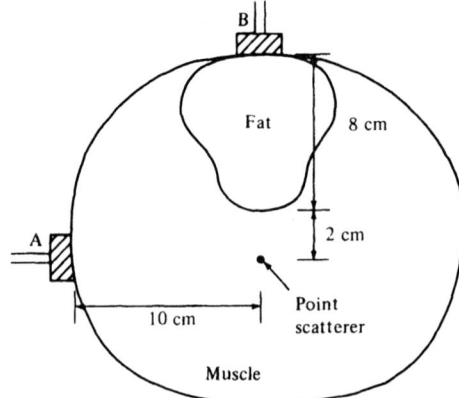

the position transducers associated with the arm introduce no error in registration, the instrument is calibrated for the phase velocity of muscle, and the display is full-scale size. Compare this error to the axial resolution of a typical medical imager.

6.8. The transducer array of a linear-array real-time imaging instrument has 32 unfocused elements. Each element is 0.5 cm wide, and there is a nonradiating gap of 0.5 cm between neighboring elements. Only one element at a time is excited.

(a) At a frequency of 3 MHz, find the lateral resolution of this instrument at a depth of 1 cm and at a depth of 8 cm. The lateral resolution of a linear-array imager is given by the beam size or the line spacing, whichever is larger.

(b) If echo information out to a depth of 12 cm is desired, calculate the minimum time required to scan the entire array.

6.9. (a) Estimate the optimum element width (b on Figure 6.14) for a linear array that will give the smallest possible beam size at a depth of 8 cm from the array. Assume the elements are excited individually and are unfocused. The frequency is 4.5 MHz. Discuss your reasoning with sketches, and list any assumptions you find necessary.

(b) If the individual elements are capable of being focused with integral lenses on each element, estimate the optimum element width to get the best lateral resolution at a depth of 8 cm from the linear array. Assume that the elements are used individually, have a gap between neighboring elements equal to 50% of the element width ($s = 1.5b$), and the frequency is 4.5 MHz. Remember that lateral resolution of a linear array is no better than the beam size *or* the element spacing, whichever is larger. Compare your answer to that of part a.

(c) Using the smaller value from either part a or part b, determine the number of elements in a linear array that is designed to have an overall view 15 cm wide.

6.10. A phased-array imager has a 16-element array. Each element is 0.15 cm wide, with a 0.07-cm gap between neighboring elements. The transducer operates at 2.25 MHz.

(a) Produce a table which lists the time delays needed for each of the 16 elements in order to steer the beam 20° off the axis. Number the elements from 1 to 16, and steer the beam in the direction of element 1. Bias the delays so that they are all positive.

(b) Produce a table which lists the time delays needed for each of the 16 elements to focus the beam to a point 8 cm from the array (assume no beam steering). Bias for positive delays only.

(c) If echo information out to a depth of 12 cm is desired, calculate the maximum number of lines possible in the sector assuming the entire sector must be obtained in less than $\frac{1}{60}$ second.

(d) Using the number of lines found in part c, calculate the spacing between lines at a depth of 8 cm if the full-sector angular width is 60°. Compare this value to the size of the smallest focused spot possible in the beam at that depth. What is this instrument's lateral resolution at that depth?

(e) If five equal-length focal zones are desired in the receiving mode out to the maximum reception depth of 12 cm, calculate how quickly the focal time delay information must be updated to the receiver analog phase shifter modules.

(f) Find the angular width of the main lobe, the position of the grating lobes, and roughly sketch the polar far-field beam pattern for this array.

6.11. It is desired to move the grating lobes of the array described in Problem 6.10 out to larger angles to reduce the grating lobe artifact. Describe one modification to the array which would accomplish this. Assume that the gap between elements cannot be made smaller than 0.07 cm. Qualitatively discuss the effect this modification would have on

(a) Lateral resolution
(b) Time delays required for a given steering angle
(c) Output power from the array
(d) System complexity and cost

6.12. A long metal needle is used to transmit ultrasonic energy from an external transducer percutaneously to a kidney stone. If the needle is carrying an average power density of 20 W/cm² at a frequency of 25 kHz, determine the peak-to-peak excursion of the tip of the needle. Assume that acoustic impedance of the metal is 4.9×10^6 g/cm² s, and neglect reflections at the tip interface.

6.13. (a) Calculate the amount of power deposited by a hyperthermia machine in a 1-cm³ volume of muscle when a 1-MHz acoustic wave of power density 10 W/cm² passes through it. Assume that scattering is small, so that absorption is given by the attenuation constant. If the specific heat of the muscle is approximately the same as water, and if heat diffusion by perfusion and conduction is neglected (an unrealistic assumption), calculate the length of time required for this volume to rise 6°C in temperature.

(b) Estimate the optimum frequency that will give the maximum power deposition at a depth of 10 cm in muscle if the power density at the surface is fixed and the beam is not focused.

6.14. For the M-mode record shown below, what is the maximum velocity of the moving object?

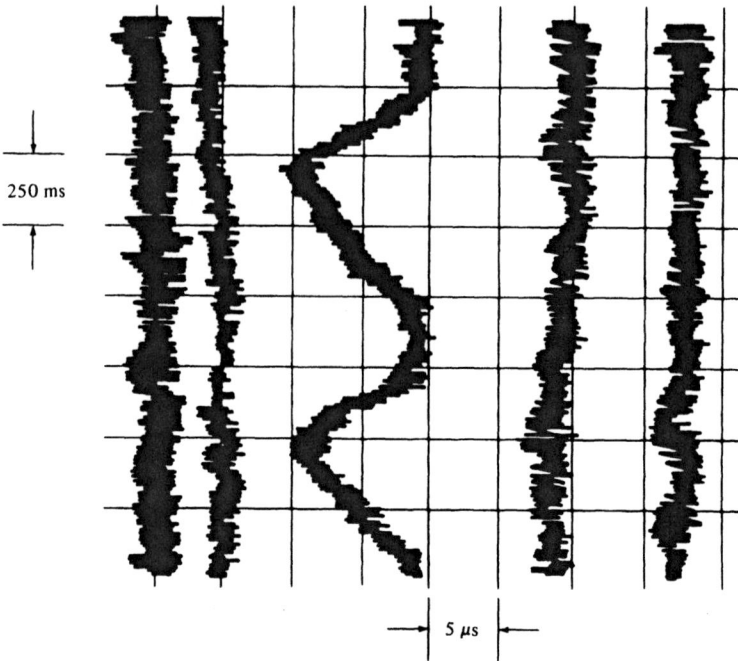

6.15. (a) Put Snell's law in terms of index of refraction, and find the indices of refraction for muscle and fat.
 (b) For the configuration shown below, find the deflection x of the beam from the straight-line path.
 (c) Calculate the time of flight for a pulse traveling the actual path. Compare this to what would be expected if water filled the space between the two transducers.

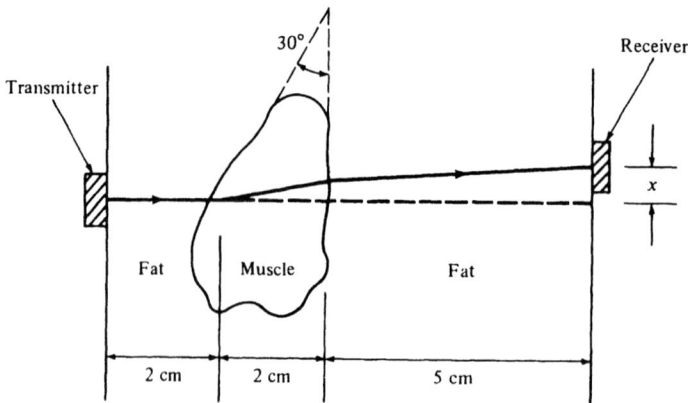

Chapter 7

Doppler and Other Ultrasonic Flowmeters

7.1 INTRODUCTION

The previous chapter described the use of ultrasound to image structures in the body. Another equally important application of ultrasound is the measurement of the velocity of moving fluids within the body, such as blood velocity in vessels or air flow in the respiratory system. The analysis of blood flow patterns is of great clinical importance in assessing cardiovascular disease.

By far the most popular means of acoustically measuring velocity relies upon the Doppler principle, whereby waves reflected from moving scatterers are shifted in frequency by an amount proportional to the velocity of the scattering objects. In the transcutaneous blood flowmeter, for example, the red blood cells flowing with the blood plasma are the moving reflectors.

Two lesser-used techniques do not require the presence of scatterers in the flowing stream. These methods, the transit time and vortex flowmeters, need only a moving stream of fluid to support the propagation of acoustic waves. Unfortunately, these methods require the placement of transducers or acoustic mirrors on both sides of the flow, which usually means an invasive cutdown when measuring blood flow. In addition, the vortex flowmeter places a small obstruction in the stream, limiting its use to external monitoring of respiratory air flow. A description of these instruments is given at the end of this chapter, after the important Doppler devices are covered.

7.2 THE DOPPLER PRINCIPLE

In Section 3.3, dealing with the reflection and transmission at stationary interfaces, it was stated that the frequencies of all the waves involved were identical. This is strictly true only for nonmoving boundaries, that is, those interfaces which have no relative motion with respect to the source and detector of the waves. In some situations, however, the reflecting boundary is moving relative to the source or receiver; in such a case the reflected wave scattered from the moving surface is shifted in frequency compared to the incident wave. This phenomenon, known as Doppler shift, has led to the development of an extremely valuable class of ultrasonic bioinstruments: the Doppler velocity flowmeters.

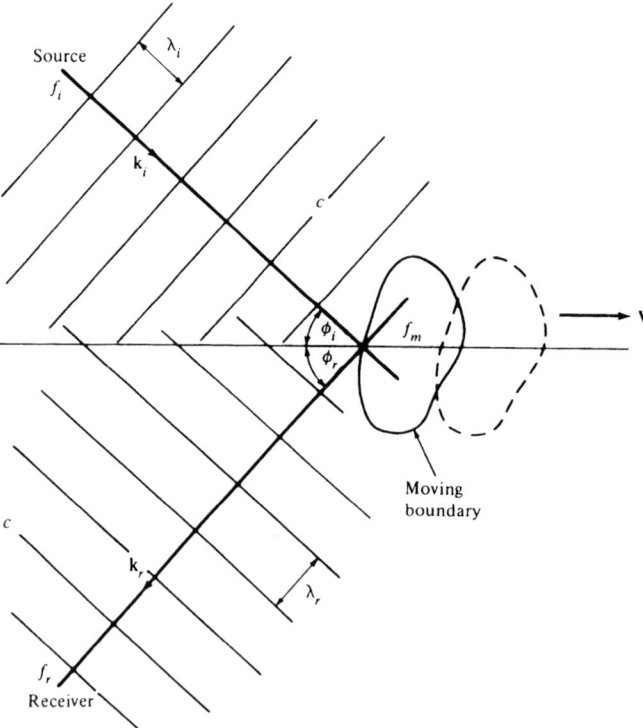

Figure 7.1 A moving boundary scattering a source wave of frequency f_i into a reflected wave of different frequency f_r. The difference $f_r - f_i$ is the Doppler shift, proportional to the scatterer's velocity V. f_m is the intermediate frequency seen by an observer on the moving scatterer. For this analysis the source, receiver, and surrounding medium are all in the same stationary frame through which the scatterer is moving.

7.2 THE DOPPLER PRINCIPLE

To derive the Doppler shift equation, consider an incident acoustic wave produced by some stationary source of frequency f_i as it encounters a moving scatterer shown in Figure 7.1. The incident wave has a wavelength λ_i in the medium, assumed stationary, surrounding the scatterer. The scatterer is moving with velocity **V** at an angle of ϕ_i with respect to the direction of propagation of the incoming wave. Upon reflection from the moving surface, the reflected wave propagates backward to the receiver situated at an angle ϕ_r with respect to the motion of the scatterer. The frequency f_r of the reflected wave as seen by the receiver (assumed to be in the same stationary frame as the source) will be different from f_i. This Doppler offset is best derived in two steps: (a) the path from source to scatterer, followed by (b) the path from scatterer to receiver.

Path from Source to Scatterer

An observer on the moving scatterer can detect the incident wave as it impinges upon the boundary, but the velocity of the wave passing over the boundary will not be the same as its phase velocity in the stationary medium because the boundary is moving away (in this case) from the source. As a consequence, the frequency f_m with which successive phase fronts are reaching the interface will be different from the sent frequency f_i. For the geometry of Figure 7.1 the incoming wave will be seen to have an effective phase velocity with respect to the moving surface given by

$$c_{in,eff} = c - V \cos \phi_i \quad (7.1)$$

Thus, the effective frequency with which the phase fronts are hitting the boundary can be found from Equation (2.27):

$$f_m = \frac{c_{in,eff}}{\lambda_i} = \frac{c - V \cos \phi_i}{\lambda_i} \quad (7.2)$$

Since in the medium $\lambda_i = c/f_i$, Equation (7.2) can be rewritten as

$$f_m = f_i \left(1 - \frac{V \cos \phi_i}{c}\right) \quad (7.3)$$

Note that the difference between f_m and f_i is proportional to the ratio of the scatterer velocity component along the wave propagation direction to the phase velocity. Also note that if either $V = 0$ or $\phi_i = 90°$, the two frequencies will be the same.

Path from Scatterer to Receiver

The impinging incident wave will set up oscillations at frequency f_m in the material of the scattering volume. These oscillations will in turn cause the reradiation of waves going away from the scatterer, one of which will be the reflected wave detected by the stationary receiver. In passing, it should be mentioned that the oscillations of the interface between two media of different acoustical impedance may be considered to be the source of all other waves derived from the incident wave, including the reflected and transmitted waves treated in Chapter 3. If the interface and the incident wave are planar, the reflected and refracted waves will be planar and will propagate in unique directions given by the law of specular reflection and Snell's law. If the boundary is irregular, the reflected and refracted waves will be scattered into many different directions.

In any case, the reflected wave at angle ϕ_r in Figure 7.1 will be radiated at frequency f_m into the medium that is moving away from the scatterer. Due to this motion, the effective phase velocity of the wave as it leaves the interface will be

$$c_{\text{out,eff}} = c + V \cos \phi_r \tag{7.4}$$

The relative motion of the scatterer will "stretch out" the wavelength of the reradiated wave in the stationary medium compared to the value it would have if the scatterer were nonmoving. This increased wavelength in the medium can be found from Equation (2.27):

$$\lambda_r = \frac{c_{\text{out,eff}}}{f_m} = \frac{c + V \cos \phi_r}{f_m} \tag{7.5}$$

As a consequence, the frequency with which this wave is detected at the receiver will be

$$f_r = \frac{c}{\lambda_r} = \frac{f_m}{1 + (V \cos \phi_r / c)} \tag{7.6}$$

Note that again the difference between the two frequencies is related to the ratio of the projected scatterer velocity to phase velocity.

Now we are in a position to combine the frequency shifts obtained over both paths into a single expression by eliminating f_m from Equations (7.3) and (7.6). This leads to a relationship between the source and received frequencies:

$$f_r = f_i \left[\frac{1 - (V \cos \phi_i / c)}{1 + (V \cos \phi_r / c)} \right] \tag{7.7}$$

A simplification may be made in Equation (7.7) by using an approximation that is valid for most practical situations:

$$V \ll c$$

7.2 THE DOPPLER PRINCIPLE

in which case Equation (7.7) reduces to (see Problem 7.1)

$$f_r = f_i \left(1 - \frac{V \cos \phi_i}{c} - \frac{V \cos \phi_r}{c}\right) \tag{7.8}$$

The Doppler shift f_d is defined as the difference between f_i and f_r according to

$$f_d = f_r - f_i \tag{7.9}$$

From Equation (7.8) we get the final result*

$$\boxed{f_d = -\frac{V}{c}(\cos \phi_i + \cos \phi_r) f_i} \tag{7.10}$$

Equation (7.10) for the Doppler shift frequency contains some interesting features. In particular, note that the shift is proportional to f_i, so the higher the source frequency, the larger the difference. However, the percentage change in f_i is independent of the value of f_i. Also note that $f_d = 0$ when $V = 0$, as is intuitively obvious.

If the direction of V is reversed such that V is in the opposite direction to that shown in Figure 7.1, then the Doppler shift will also be reversed in sign. In general, it can be said that if the reflector is moving away from the source and receiver, the reflected wave will be down-shifted in frequency; or, in other words, f_r is less than f_i so f_d is negative. If the object is moving toward the source and receiver, f_r will be larger than f_i so f_d is positive.

7.2.1 Vector Formulation

There are some situations in which f_d equals zero even though V is nonzero, depending upon the angles ϕ_i and ϕ_r. Examination of Equation (7.10) and Figure 7.1 shows that when the incident and reflected angles are in certain quadrants in relationship to the direction of V, it is possible for $\cos \phi_i$ and $\cos \phi_r$ to be equal but opposite in sign so they cancel, leading to a zero Doppler shift. Perhaps the easiest way to visualize this characteristic is to

---

*For the case where both the medium and the scatterer are moving with velocity V with respect to the source and receiver, as is the case with red blood cells immersed in a moving stream of blood, Equation (7.7) will be replaced with

$$f_r = f_i \left[\frac{1 - (V \cos \phi_r/c)}{1 + (V \cos \phi_i/c)}\right]$$

as shown in Problem 7.2. However, using the approximation that $V \ll c$, this equation reduces to Equation (7.8) so both configurations lead to the same final result for the Doppler shift frequency.

put the Doppler shift formula into vector form. To do this, define a propagation vector **k** associated with each wave whose magnitude is the propagation constant $k = 2\pi/\lambda$ used in previous chapters and whose direction is the direction of propagation of the wave (perpendicular to the planar phase fronts for plane waves). Let **V** be a vector in the direction of the motion of the scatterer with a magnitude V. Then, Equation (7.7) can be rewritten in a much simplified and easy-to-remember form by using vector dot products. Since the dot product of two vectors **a** and **b** is defined by **a** • **b** = $ab \cos \theta$, where θ is the angle between the two vectors, Equation (7.7) becomes

$$f_r\left(1 - \frac{\mathbf{V} \cdot \mathbf{k}_r}{ck_r}\right) = f_i\left(1 - \frac{\mathbf{V} \cdot \mathbf{k}_i}{ck_i}\right) \tag{7.11}$$

Remembering that $c = \omega/k$ and that $\omega = 2\pi f$, Equation (7.11) can be rewritten as

$$\omega_r - \mathbf{V} \cdot \mathbf{k}_r = \omega_i - \mathbf{V} \cdot \mathbf{k}_i \tag{7.12}$$

Since $\omega_d = \omega_r - \omega_i$, Equation (7.12) leads to a very compact vector equation for the angular Doppler frequency ω_d:

$$\boxed{\omega_d = \mathbf{V} \cdot (\mathbf{k}_r - \mathbf{k}_i)} \tag{7.13}$$

Equation (7.13) has employed no approximations to this point, so it is exact. However, it has one awkward feature. On the right-hand side the values of both frequencies ω_i and ω_r enter because $k_i = \omega_i/c$ and $k_r = \omega_r/c$. Yet only ω_i is known initially. To simplify the use of Equation (7.13), it is usually assumed that since $V \ll c$, $\omega_d \ll \omega_i$ and $\omega_r \approx \omega_i$. Thus, the *magnitude* of the vector $\mathbf{k}_r$ is usually considered to be equal to the magnitude of $\mathbf{k}_i$:

$$k_r \approx k_i \tag{7.14}$$

The *directions* of the two vectors can be very different, however, so the resultant vector appearing in Equation (7.13), $\mathbf{k}_r - \mathbf{k}_i$, can take on a large range of values and directions. This resultant vector, which we shall call the "sensitivity" vector of the particular configuration being considered, is very important because its magnitude and direction determine the sensitivity of the Doppler instrument to the scatterer's velocity **V**.

Equation (7.13) can be used to calculate the Doppler shift ω_d for any situation by the following steps:

1. Use the direction of incidence and the incident frequency to determine $\mathbf{k}_i$.
2. Reverse the vector to get $-\mathbf{k}_i$.

7.2 THE DOPPLER PRINCIPLE

3. Use the direction of reception and the magnitude k_i to determine $\mathbf{k}_r$.
4. Vectorially add $\mathbf{k}_r$ to $-\mathbf{k}_i$ to yield the resultant sensitivity vector $(\mathbf{k}_r - \mathbf{k}_i)$.
5. Take the dot product of the sensitivity vector with $\mathbf{V}$ to determine ω_d.

7.2.2 Examples of Doppler Configurations

Some examples of various configurations of source and receiver positions and the relative sensitivities of these arrangements for measuring the velocity V are shown in Figure 7.2. A comparison of the four different configurations shown in the figure gives some idea of the unique characteristics that each possesses for measuring the scatterer velocity. These configurations, for instance, might be considered to be four possible ways of positioning the source and receiver transducers in an instrument to measure blood velocity transcutaneously via Doppler scattering from moving red blood cells. Figure 7.2a shows an arrangement for which the resultant sensitivity vector $(\mathbf{k}_r - \mathbf{k}_i)$ is longitudinal, so the system is sensitive only to the longitudinal flow of the blood; the dot product of the assumed velocity vector $\mathbf{V}$ and $(\mathbf{k}_r - \mathbf{k}_i)$ is a maximum negative value since the angle between these vectors is 180°. By the same token, the system is insensitive to any transverse velocity components, such as scattering from a nearby vessel wall as it pulsates radially during each heart cycle, since these components will have a vector direction perpendicular to $(\mathbf{k}_r - \mathbf{k}_i)$ and their corresponding dot product is zero. The fact that the Doppler shift is negative is caused by the antiparallel nature of $\mathbf{V}$ and $(\mathbf{k}_r - \mathbf{k}_i)$; that is, the source frequency is downshifted because the scatterer is moving away from the source and receiver. This negative value presents no difficulty for the proposed system. Many ultrasound Doppler instruments use a detector which mixes a portion of the source wave with the received signal in a nonlinear device such as a diode to obtain the difference frequency. Unless specially designed, these instruments do not detect the sign of the Doppler shift. When the direction of flow is important, bidirectional flowmeters have been developed that give an indication of the flow direction as well as its velocity; these devices are discussed later in this chapter.

In spite of the relatively large sensitivity to longitudinal flow possessed by arrangement a in Figure 7.2, it suffers one practical defect as a possible blood flow instrument. It requires that the receiver be positioned on the opposite side of the flow from the source transducer, which is impractical for most situations where intervening tissue, bone, or air masses limit the field of view of the blood vessel to be from one side only.

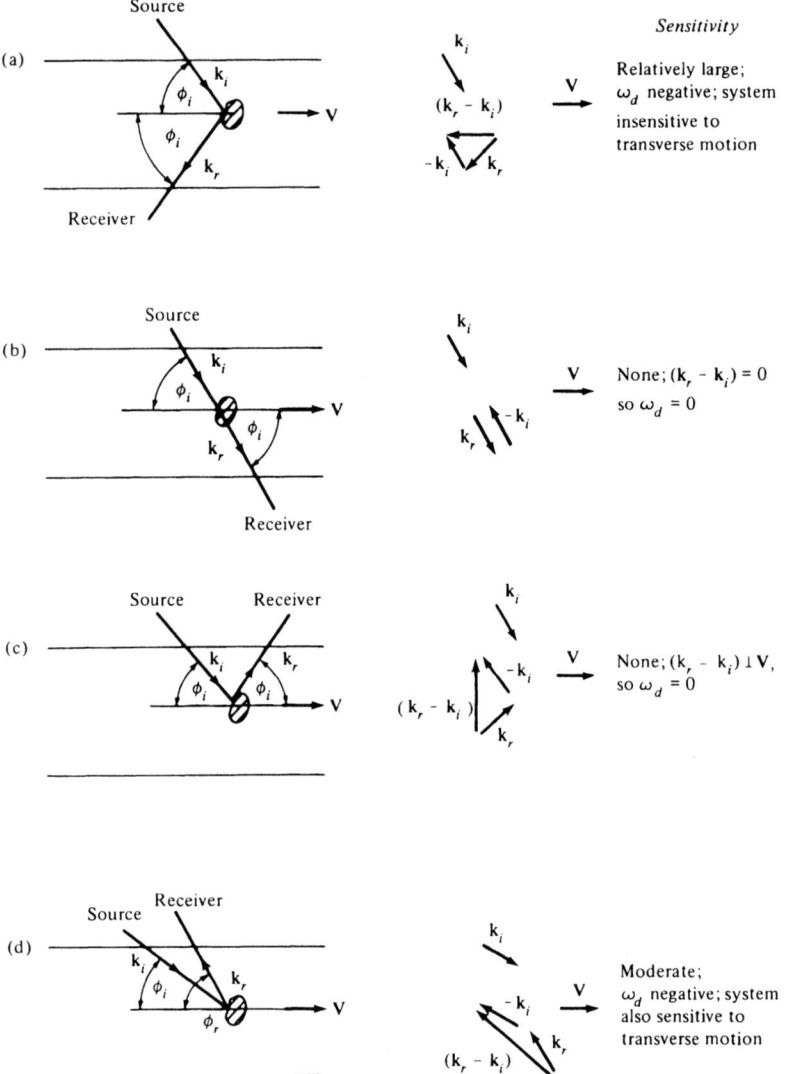

Figure 7.2 Some configurations of different source and receiver positions relative to the direction of velocity of the scatterer for a Doppler velocimetry system. The sensitivity of the system to measure **V** may be obtained from Equation (7.13).

An examination of the configuration in Figure 7.2b shows that not only does it require the source and receiver to be on opposite sides of the flow, but that if they are arranged in a straight line with the scattering point, there is no Doppler shift whatsoever since $\mathbf{k}_r$ and $-\mathbf{k}_i$ cancel each other. Even if the source and receiver are not precisely in line with the scatterer, the vector ($\mathbf{k}_r - \mathbf{k}_i$) will still be small, leading to very low sensitivity to velocity. So configuration b normally is not practical either.

Configuration c would seem to be a likely candidate for a blood flowmeter, since the source and receiver are now on a common side, but examination shows that the resultant sensitivity vector ($\mathbf{k}_r - \mathbf{k}_i$) is now in the transverse direction, so the instrument is sensitive only to transverse velocity and not sensitive to the longitudinal velocity V of the blood. By making the angle of reception slightly different from the angle of incidence, the resultant vector will have a small longitudinal component for detecting longitudinal flow, but it will be small compared to the overwhelming sensitivity to transverse velocity.

Configuration d in Figure 7.2 overcomes many of the previous objections by placing the receiver and source close together with a fairly shallow angle between them and the velocity vector V. This gives a sensitivity vector ($\mathbf{k}_r - \mathbf{k}_i$) which, although not precisely longitudinal, still has a sizable longitudinal component. There is some sensitivity to transverse motion, such as reflections from the outwardly pulsating vessel walls in an artery during systole, but these signals may often be discriminated from the blood velocity signals on the basis of their generally lower frequency content. Note that the source and receiver are on the same side of the flow; in fact, some instruments (notably pulsed Doppler systems) are able to combine the two into one transducer element.

Angles

In all of the arrangements shown, it is assumed that the angles ϕ_i and ϕ_r are known in order to quantitatively determine f_d from either Equation (7.10) or Equation (7.13). In most cases, however, the vessel carrying the blood is hidden underneath overlying tissue, and its course cannot be readily determined. This leads to uncertainty in the angles and some error in the measurement of the magnitude of V and represents one of the challenges for Doppler blood velocimeters. Clever ways of determining these angles is the subject of much interest among instrument designers and users, and it will be further discussed in the section on pulsed Doppler flowmeters.

In the absence of precise angular information, many hand-held Doppler scanners are calibrated assuming a tilt angle of 45° with respect to the blood vessel axis (assumed parallel to the skin surface) as sketched in Figure 7.3a. This angle is relatively easy for the operator to approximate and represents a tradeoff between too large an angle (Figure 7.3b), which results in low-frequency shifts for longitudinal velocity components compared to transverse motion, and too small an angle (Figure 7.3c), which leads to long path lengths and high attenuation through the intervening tissue. Problem 7.4 shows that the magnitude of the error caused by inadvertent angular mispositioning of the transducers is not too severe.

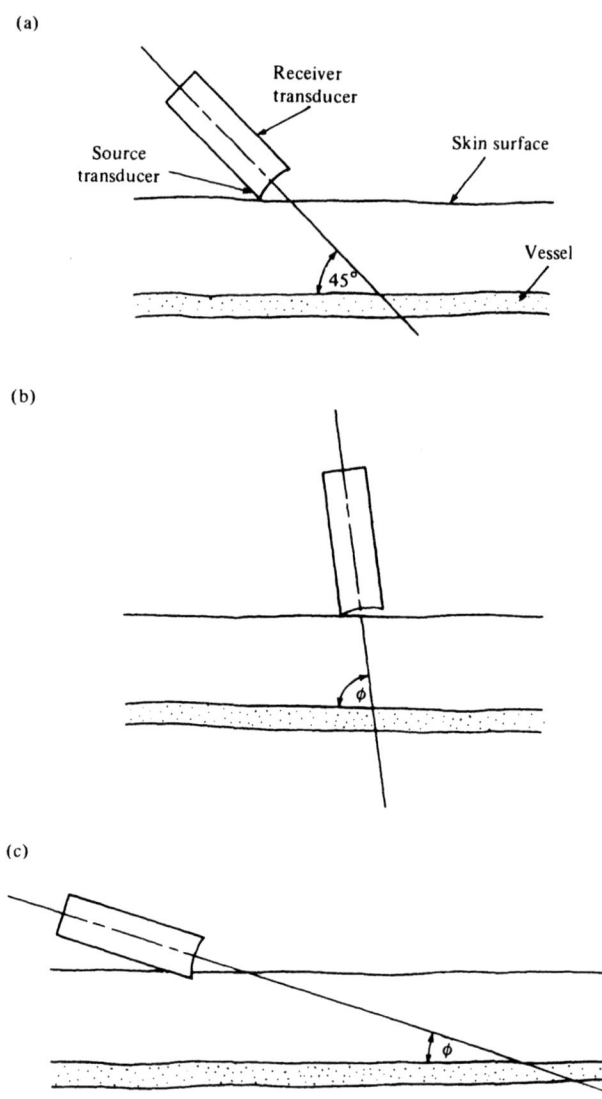

Figure 7.3 (a) Many hand-held Doppler units are designed to be held at an angle of about 45° with respect to the skin's surface. (b) If the angle is too near perpendicular, little frequency shift will be seen. (c) If the angle is too shallow, long intervening tissue path lengths will cause high attenuation.

Spectral Spread

The equations developed so far in this section have assumed a single value for the Doppler frequency. In actual practice of blood flow measurement,

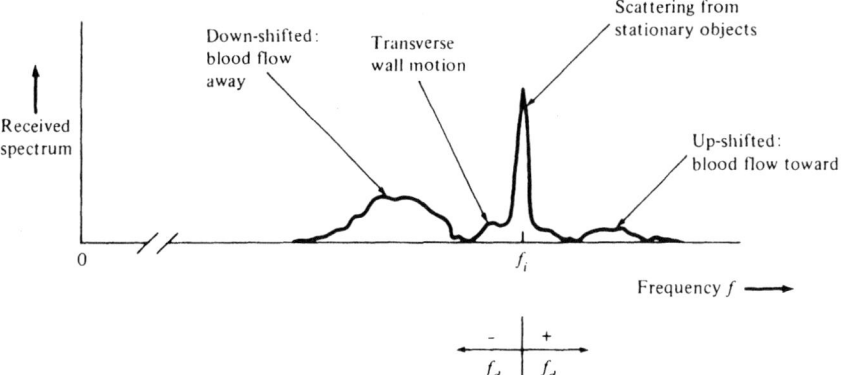

Figure 7.4 The frequency spectrum of ultrasound scattered by an artery which is predominantly carrying blood away from the transmitter/receiver position. The Doppler-shifted spectrum is broadened by the blood velocity profile, variation in beam angle, and the finite viewing time for each scatterer. The low frequency Doppler returns are due to transverse motion of the artery's walls. Note that since $f_i \gg f_d$ for blood flow, the frequency scale has been broken to allow an expanded view of the Doppler spectrum.

the Doppler return will be a moderately broad spectrum of frequencies whose distribution is given by applying the equations to each segment of the flow. For example, the finite beam widths of the incident and received beams produce a finite *volume* of overlap where the scatterers are illuminated and viewed, rather than a single point. Spatial and temporal variations in blood velocity profile within this volume, as well as variations in the incident and received angles across the volume, will lead to a spread of the spectrum, as shown in Figure 7.4.

Attempts to reduce the sampling volume size will lead to a less representative sample, and scanning must then be implemented to yield a complete profile of the blood velocity. If the sampling volume becomes too small, the window in time during which each scatterer is seen (t' = sample distance/velocity) will become so small that the spectrum will be spread due to the Fourier inverse relationship $\Delta f \approx 1/t'$ describing the frequency-broadening effect of a small time aperture (refer to Section 6.2.1 and Figure 6.5).

7.3 CW DOPPLER FLOWMETERS

In its basic form the transmitter of the Doppler flowmeter is excited with a continuous sinusoidal voltage and the signal waveform from the receiver transducer is mixed with a portion of the transmitted waveform in a nonlinear device such as a diode. This arrangement is diagrammed in Figure 7.5.

The frequency content of the output from the diode mixer can be found by a consideration of the general results of *mixing theory,* in which two sinusoidal waveforms of the same or different frequencies are first added and then passed through a nonlinear device. The output of the mixer will contain several sinusoidal components representing all possible sum and difference frequency combinations of the original frequencies (see Problem 7.6). For our purposes here, only the difference-frequency component is of value and is retained.

For example, if the two input frequencies are f_i (transmit frequency) and $f_i + f_d$ (Doppler-shifted return), the difference-frequency component after mixing will be at a frequency f_d. This holds just as well for a spectrum of various f_d values as for a single f_d. Therefore, by the process of mixing with a signal at the transmitted frequency, we have extracted the Doppler frequency spectrum desired.

A general way of stating the principle of mixing is to note that when a received signal possesses a spectrum centered around frequency f_i and it is mixed with a reference signal at single frequency f_{LO} (the local oscillator), the result after mixing is to shift the signal spectrum down and center it around a new intermediate frequency $f_{if} = f_i - f_{LO}$. The new center frequency is a result of the difference-frequency component in the mixing process, called heterodyning for this case. Any portions of the shifted spectrum which happen to fall in the negative frequency region are "wrapped around" onto the positive frequency side, since $\cos(-2\pi ft) = \cos(2\pi ft)$, where they combine with the positive frequency portion by appropriate phase angles.

Applying this principle to the flowmeter shown in Figure 7.5, the original unmixed receiver spectrum is similar to Figure 7.4 and is centered about f_i. After mixing with a portion of the transmitter signal* at frequency $f_{LO} = f_i$, the result will be centered at $f_{if} = f_i - f_{LO} = f_i - f_i = 0$ (that is, down around dc). This shifted spectrum is shown in Figure 7.6. Note that since any negative-frequency spectral components are wrapped around onto the positive-frequency side, this detection scheme as it stands cannot differentiate between down-shifted and up-shifted Doppler returns, that is, it cannot determine the direction of the blood flow.

The resultant spectrum in Figure 7.6 contains some unwanted terms: the zero- and low-frequency clutter from stationary tissues and transverse wall pulsations, and all the sum-frequency terms (not shown) due to the nonlinear mixing. The desired blood flow spectrum can be isolated by

---

* Injecting the transmitter signal as a separate input to the mixer is often not necessary in practice since the frequency f_i is already unavoidably present in the return from miscellaneous stationary objects.

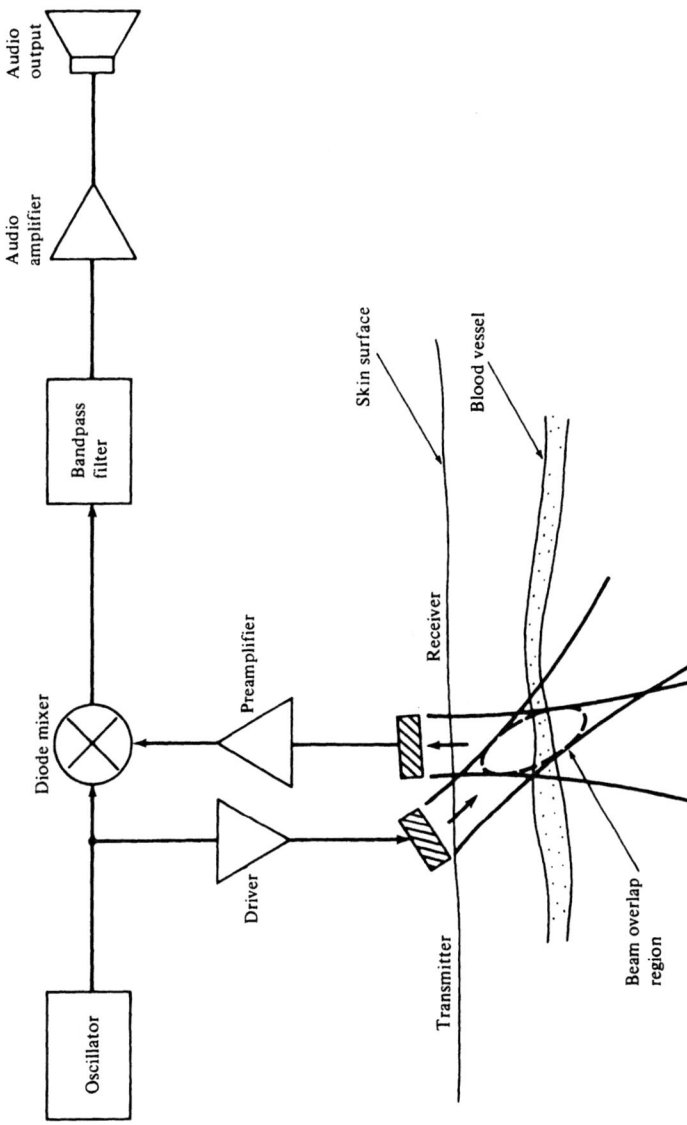

Figure 7.5 A simple continuous wave (cw) Doppler flowmeter whose output is an audible signal from a loudspeaker; earphones are also sometimes used. The velocity spectrum of the blood cells appears as an audible spectrum of sound.

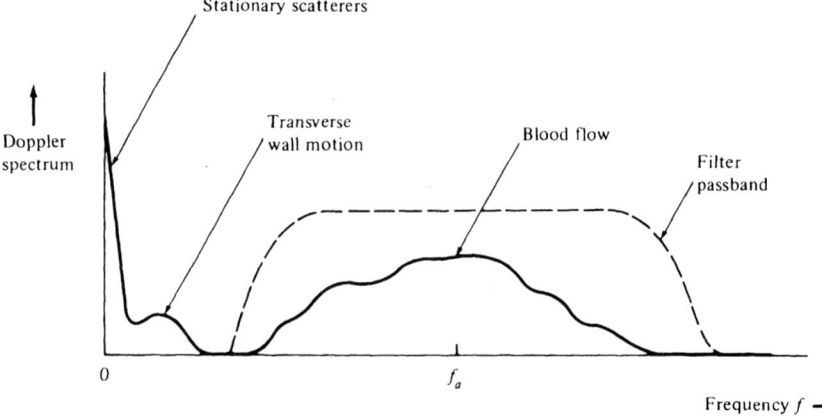

Figure 7.6 Typical spectrum of received Doppler frequencies from an artery near the time of peak flow. By mixing with f_i the Doppler spectrum is centered around $f = 0$. The signals from nonmoving boundaries and pulsating walls can be largely removed by bandpass filtering. f_a is the mean frequency of the blood flow portion of the spectrum and is representative of mean blood velocity. At other times in the heart cycle, all components will be scaled downward owing to a lower blood velocity.

passing the signal through a bandpass filter positioned around the expected Doppler spectrum. The lower cutoff frequency characteristic of the filter (usually at 30 to 100 Hz) should be as sharp as possible to avoid attenuating low-velocity blood flow signals while still rejecting wall motion artifacts; see Problem 7.7.

After filtering, the type of processing done next on the signal will determine the eventual form of the instrument's output. There are several options, discussed below in order of their increasing complexity.

7.3.1 Audible-Output Instruments

If the filtered spectrum is simply amplified by an audio amplifier and fed to a loudspeaker or set of earphones, the Doppler information is presented as audible sound to the user. This is due to a fortunate circumstance: The range of Doppler frequencies for most flowrates in the body, when the incident frequency is in the 2- to 10-MHz range, falls within the audible portion of the sound spectrum, from about 20 Hz to approximately 10 kHz (see Problem 7.7).

When placed over an artery, the Doppler spectrum from the loudspeaker has a "whooshing" or "swishing" sound which peaks during systole and drops in frequency during diastole. The variation in frequency heard during a heart cycle is related to the change in blood velocity in the artery.

In the veins the sound spectrum is much lower and more constant, like a low roar, due to the slower, steadier flow in the veins. The human hearing apparatus is quite good at performing a frequency analysis of incoming sounds, and expert users become adept at detecting subtle changes in the spectrum related to the state of the underlying vessel.

This audible-output Doppler device is small, portable, and simple to use. One key application is scanning peripheral blood vessels to spot regions of partial or complete occlusion. When a vessel becomes constricted over a short portion of its length, the law of conservation of volumetric flow dictates that the velocity of the blood must locally increase as it squeezes through the stenotic region. This can be detected as an increase in the frequency of the Doppler "whoosh" when the instrument passes over the site of narrowing. In more severe cases of occlusion the lack of any detectable Doppler return in a region may indicate loss of blood flow to that area. In this fashion arterial and venous pathways may be tracked and mapped.

Also, when blood pressure readings are taken with an inflatable cuff, the presence of the Doppler sound is useful as indicator of barely released arterial flow ("systolic breakthrough"). The onset of the sound is a more reliable systolic pressure indicator for vessels too deep or too small (in neonates) to be accurately judged by palpation.

In obstetrics, lower-frequency Doppler devices will image the pregnant uterus to monitor fetal life, blood flow, and heart rate. Doppler instruments are also valuable tools for assessing heart valve motion in cardiac patients.

Frequency Choice

The choice of the optimum frequency for a desired depth is governed by two conflicting factors: (1) For good penetration without high absorption, a lower frequency is best since tissue attenuation increases roughly proportionally to f. (2) For maximum scattered power from the collection of red blood cells, a higher frequency is best since each cell scatters by an amount that is approximately proportional to f^4, as explained next.

The scattering behavior of small particles depends upon the ratio of particle size d to the wavelength λ of the incident radiation. When $d \ll \lambda$, as is the case for the red blood cell (see Problem 4.2), the particle's scattering may be classified as falling in the "small-particle," or *Rayleigh,* regime. The general characteristics of Rayleigh scattering are as follows: (1) the scattering cross section varies as f^4; (2) the scattering direction possesses a uniform angular distribution (isotropic reradiation); and (3) there is a general insensitivity to the exact shape or orientation of the scatterer.

When the rapid variation of scattered power with frequency is balanced against the frequency dependence of attenuation in intervening tissue, an optimum frequency for Doppler as a function of depth may be determined. For shallow vessels near the surface of the skin, an incident frequency of between 4 to 10 MHz is usually chosen. For deeper penetration in abdominal probing, the incident frequency is lowered to near 2 MHz (see Problem 7.8).

7.3.2 Analog-Output Instruments

The audio output of the above instruments provides qualitative information to the user. For more quantitative measurements, electronic signal processing can be done on the Doppler spectrum after mixing and filtering and an analog output voltage supplied for recording and display. For example, the filtered signal may be passed through a *zero-crossing detector,* which electronically detects the frequency with which the waveform passes through the zero voltage level. This frequency is loosely related to the mean frequency of the Doppler spectrum, depending upon the exact shape of the spectrum and the amount of hysteresis built into the zero-crossing detector to avoid sensitivity to noise; see Figure 7.7. In practice, the zero-crossing frequency falls somewhere above the average frequency value and below the peak frequency component; the relationship will change for different portions of the heart cycle and for various hysteresis settings.

An analog signal is then generated proportional to the zero-crossing frequency, indicating quasi-instantaneous blood velocity, and this signal may be displayed on a CRT screen or recorded on a stripchart.

Bidirectional Flowmeters

As mentioned, the simple mixer in Figure 7.5 will not differentiate between forward and reverse flows. Reversal of flow direction can occur within a

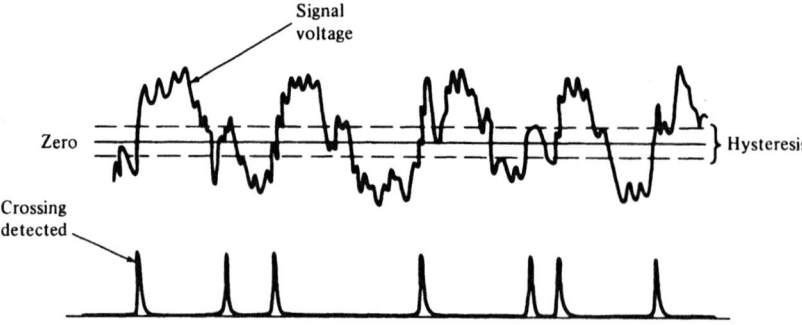

Figure 7.7 A zero-crossing detector counts the rate at which the signal crosses over (in one direction) a region near zero voltage. To avoid oversensitivity to noise fluctuations, a hysteresis band is usually included; to be counted as a crossing, the signal must pass through the entire band.

7.3 CW DOPPLER FLOWMETERS

single vessel at different phases of the heart cycle or even simultaneously at different radii of the vessel during the transition period from systole to diastole. Sometimes, the instrument's overlapping beams include both an artery and a vein in the same view, resulting in scattering from counter-directed flows.

To avoid this ambiguity in direction, bidirectional flowmeters have been developed that are able to distinguish between reverse and forward flow. Recall that the direction of flow is coded in the original frequency spectrum of the returned signal by whether the received frequencies are Doppler-shifted higher or lower than f_i. This spectral relationship is lost by wrap-around if the Doppler spectrum is mixed down to zero frequency, as occurs when the transmitter signal at f_i is used in the mixing (Figure 7.6). But the directional relationship can be preserved if the Doppler spectrum is not shifted all the way to zero. For example, one bidirectional scheme uses a local oscillator mixing frequency which is offset from f_i by an amount f_{if} (i.e., $f_{LO} = f_i - f_{if}$) where f_{if} is much smaller than f_i but is still higher than the largest expected Doppler shift frequency. The result after mixing is a Doppler spectrum that is centered around f_{if}, not zero, and has its up-shifted and down-shifted portions still distinct; see Figure 7.8. These two portions of the spectrum can then be separated by two bandpass filters before detection (by two zero-crossing detectors, for example) to give velocity output waveforms for either direction when flow is present in that direction. The bandpass filters used to split the spectrum unfortunately need very sharp cutoff characteristics on either side of f_{if} to avoid clipping low-velocity flow signals while still rejecting the stationary and wall-motion clutter.

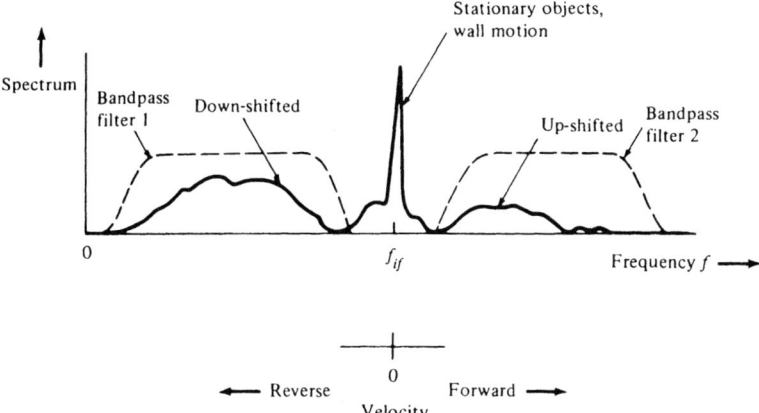

Figure 7.8 The Doppler spectrum after its original center frequency f_i has been shifted to an intermediate frequency f_{if} by mixing with frequency $f_{LO} = f_i - f_{if}$. The spectral portions on either side of f_{if} indicate the amount of flow in each direction; one side usually predominates at any given time.

A widely used alternative method is called *quadrature-phase detection*. In this technique a signal 90° out of phase (in quadrature) with the transmitting signal but at the same frequency is generated, and the Doppler-shifted return is mixed with both the original transmitting signal and the quadrature signal to form two channels. The phase relationship (lead or lag) between these two channels determines whether the flow is forward or reverse.

Bidirectional Doppler instruments have found extensive use in analyzing peripheral vascular disease and in scanning the arteries of the head and neck such as the carotid arteries. In these procedures the shape of the velocity waveform is carefully studied, looking for modifications to the systolic and diastolic portions of the waveform which are characteristic of occlusions, stiffened arterial walls, or collateral circulation. The abnormal waveform may even be computer-matched to a data bank of waveforms which are representative of various arterial disease states.

7.3.3 Spectrum-Output Instruments

To gain a more complete picture of the flow pattern than can be provided by a single analog output waveform, some newer instruments display the frequency spectrum of the returned signal by performing a Fourier transform operation by such means as the fast-Fourier transform (FFT) on a digital computer or by a parallel arrangement of sequentially-spaced band-pass filters. This gives the user a view of the width, position, and shape of the entire spectrum and shows the dynamics of the return during a heart cycle. The spectral shape may be something like Figure 7.6 or Figure 7.8 depending upon the mixing technique employed. Once the spectrum is found, it is of value to calculate and display flow parameters such as mean flow velocity, peak velocity, and velocity spread.

Figure 7.9 shows the spectrum display of one doppler instrument used for visualizing the flow in the carotid arteries and other major vessels. The vertical axis of the display is proportional to velocity (Doppler frequency), including a region below the axis for reverse flow. The horizontal axis shows time during one or more heart cycles. The amplitude of the spectrum is displayed as changes in color on the screen, progressing from black for no energy through shades of red and yellow to white for maximum energy.

7.4 DOPPLER IMAGERS

The cw Doppler flowmeters described above provide only partial spatial information regarding the positions of the moving scatterers, deduced from

7.4 DOPPLER IMAGERS

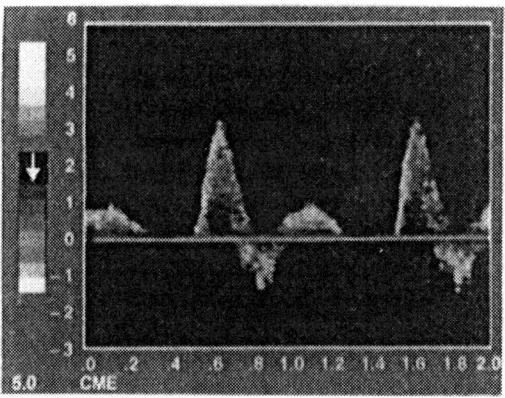

Figure 7.9 (a) Example of the spectrum display of a Doppler instrument viewing the flow in a normal common femoral artery. The vertical axis is proportional to Doppler shift frequency (reverse flow is displayed below the zero position) while the horizontal axis shows the time history during two heart cycles. The magnitude of the spectrum in this display (originally colored) is coded as a progression from black (no energy) through shades of red and yellow to white (highest energy). This black-and-white reproduction loses some of the impact of the actual color display. (b) This instrument is also capable of B-scanning and pulsed Doppler range gating. Photographs courtesy of Carolina Medical Electronics, Inc.

the location of the overlapping beam patterns of the transmit and receive transducers. Since the two transducers are usually aimed in nearly the same direction with only a shallow angle separating them (see Figure 7.2d), the overlap volume and consequent position ambiguity can be large.

There are many situations, however, where more precise spatial resolution would be of significant clinical value. For example, at an arterial branch or other site where more than one vessel may be in the field of view at the same time, it is important to spatially separate the individual flow contributions. In addition, in cases of suspected stenosis, a spatial image of the laterally narrowed flow combined with the Doppler frequencies helps to confirm the diagnosis. Also, if images are obtained of the blood vessel and its orientation, the angular uncertainty entering Equations (7.10) and (7.13) can be substantially reduced.

Finally, Doppler frequency is basically a measure of blood velocity, not volumetric flow. To convert velocity to volume flow rates, the vessel's cross-sectional area (or diameter if assumed circular) as well as the velocity profile across the vessel must be obtained from spatial data.

Several instruments are described next which have been developed to provide some spatial information regarding the Doppler returns. This is currently one of the most rapidly growing areas of ultrasonic instrumentation development.

7.4.1 CW Transverse Scanners

This device is essentially a cw Doppler transducer rigidly attached to an articulating arm which allows the head to be scanned transversely over the region imaged while being held at a fixed angle to the skin's surface. Electronic sensors on the arm measure the location of the head in the x–y plane parallel to the skin's surface, and the display presents the magnitude of the Doppler frequency return in B-mode fashion on the screen. By manually passing the head back and forth over the area, a two-dimensional pattern of velocity is built up. The velocity magnitude is presented either as an increasing brightness at each corresponding spot of a black-and-white display, or as a color code on a color monitor (for example, black = no flow, red shades = low flow, yellow shades = medium flow, and white = high flow).

The spatial resolution in the axial direction (depth into the body surface) is not improved over a conventional cw Doppler device since it is determined by the long beam overlap volume, which must accomodate a range of possible scattering depths. The resolution in the lateral direction, though, may be quite good since narrow, focused beams are employed. Therefore, these instruments work best when used to image surface vessels whose branching has primary components in a plane parallel to the skin's surface. This is the case for some regions of the major cerebral vessels passing through the neck: the common, external, and internal carotid arteries, and the jugular vein. Here, ultrasonic views can yield information regarding patency and blood flow in cases of suspected stenosis or occlusion.

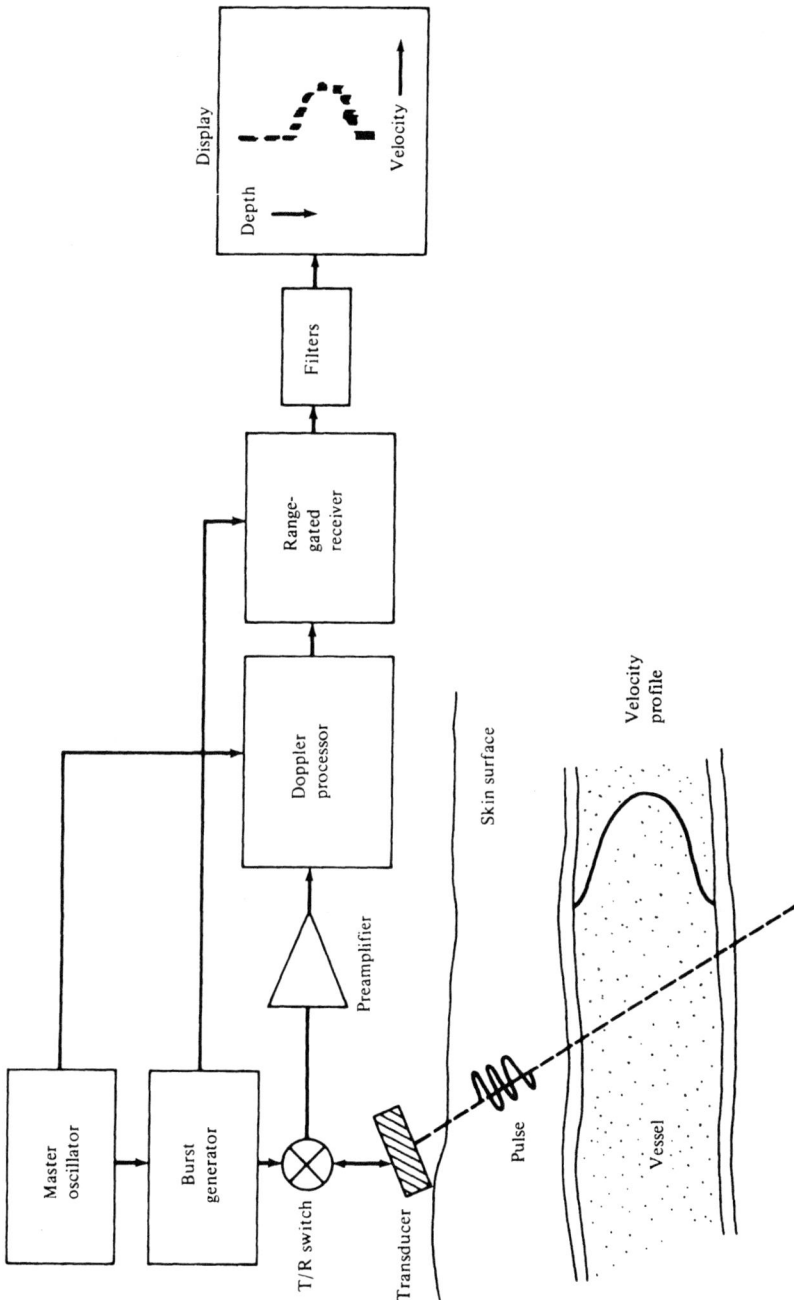

Figure 7.10 The pulsed Doppler flowmeter correlates the dimension of depth with the Doppler-shifted echoes, allowing velocity profiles and vessel size and location to be displayed.

7.4.2 Pulsed Doppler Flowmeters

To give improved resolution in the axial direction, pulses of incident energy are used, and detecting the arrival time in combination with the Doppler shift of the echoes (reminiscent of A-mode) allows the velocity data to be correlated with depth of penetration. Figure 7.10 shows the basic arrangement. Pulses a few cycles long of incident frequency f_i are produced by a burst generator driven from a master oscillator. As each transmitted pulse

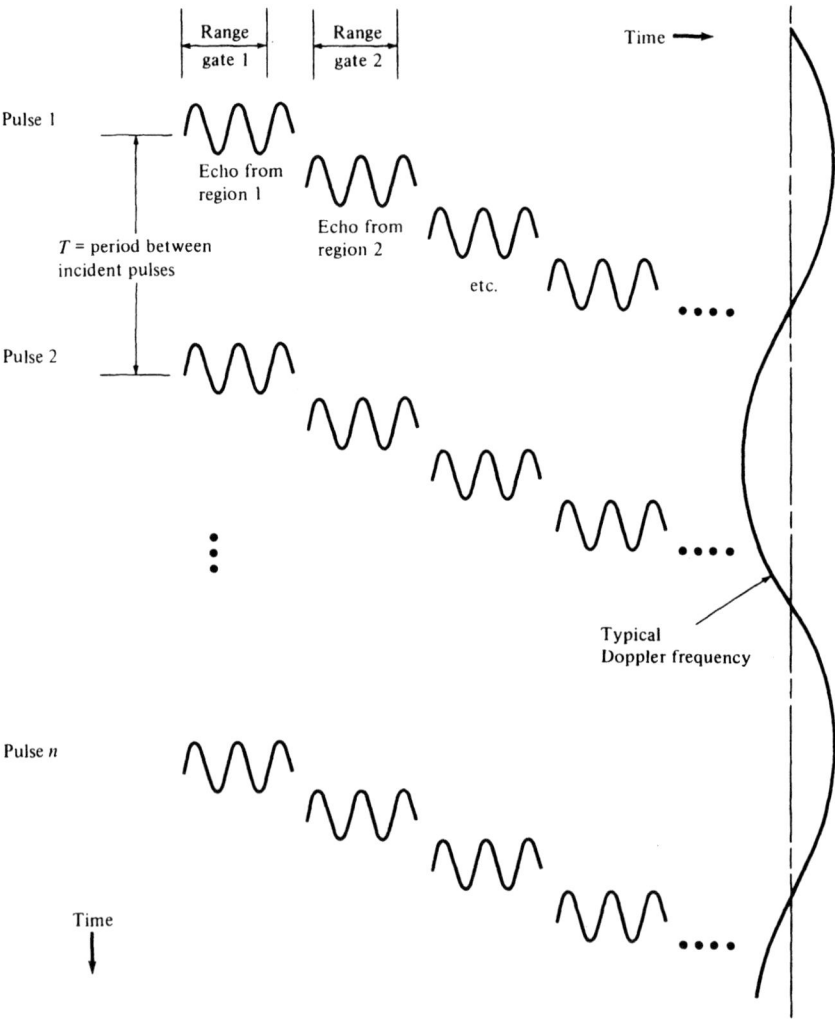

Figure 7.11 Timing relationship between incident pulses, range-gated returns, and the period of a typical Doppler shift.

7.4 DOPPLER IMAGERS

traverses the imaged region, echoes are reflected whose frequency is Doppler-shifted proportional to the velocity of the scatterers encountered. After preamplification, this Doppler information is processed by a means described in greater detail later. The receiver stage is triggered such that it is sensitive to echoes only during a limited time window corresponding to a specified depth of pulse penetration; this is known as "range-gating" the receiver. The magnitude of the Doppler shift is displayed as a velocity value at the appropriate depth position on the screen.

Due to the low ratio of blood velocity to speed of sound (V/c), the Doppler shift f_d is only a small percentage of the incident frequency f_i. Thus, during the short length of each pulse required to give good range resolution (a few cycles of f_i), the change in frequency in the echo is almost imperceptible. It is therefore necessary to use many pulses at each position of the range gate to build up enough information to accurately reconstruct the value of f_d at that range.

This is shown in Figure 7.11. Each pulse (for example, pulse 1) produces echoes which are compartmentalized into receiver range gates corresponding to the depth of their reflection. At each range position, several successive incident pulses must be accumulated to accurately trace the waveform of the Doppler frequency shift, shown in the right margin of the figure.

A more precise way of stating this requirement is to note that a single echo will contain a very broad range of frequencies since it must be short to give good axial resolution. From the discussion of Fourier relationships which led to Figure 6.5, a sinusoidal pulse of length t' will be broadened into a frequency spectrum whose width is approximately given by

$$\Delta f \approx \frac{1}{t'} \qquad (7.15)$$

As a numerical example, let $f_i = 5$ MHz so $\lambda = 0.3$ mm. To achieve an axial resolution of 1 mm, the incident pulse must be only 6.7 cycles long by Equation (5.24); thus $t' = 1.3$ μs. The frequency of the pulse is broadened (and therefore becomes uncertain in precise frequency) by an amount $\Delta f = 1/(1.3$ μs$) = 770$ kHz. In contrast, the Doppler shift will be much less than 10 kHz at even its highest peak, and it therefore will be hidden in the width of the spectrum of this single pulse. However, if many samples are collected over a time that contains several periods of the Doppler frequency, the effective time width is dramatically lengthened, allowing the Doppler frequency to be accurately determined.

The spacing T between incident pulses is determined by the pulse repetition frequency (PRF) selected. The lower limit to this rate is set by the Nyquist sampling criterion, which states that to adequately determine a waveform of frequency f_d it must be sampled at a rate at least as high as $2f_d$ (that is, at least two samples per period) in order to reconstruct the waveform without ambiguity or aliasing effects. Thus, if the highest Doppler frequency expected is f_{dh}, the time between pulses can be no longer than $T = 1/2f_{dh}$. This in turn determines how deep the instrument can image since echoes from each pulse must all be returned from the farthest depth (denoted l_m) before the next pulse can be sent. Thus, T can be no shorter than $2l_m/c$.

There is therefore a tradeoff between the highest detectable Doppler frequency f_{dh} and depth of penetration l_m for unambiguous echoes. From the discussion above,

$$\frac{2l_m}{c} \leq T \leq \frac{1}{2f_{dh}}$$

or

$$l_m \leq \frac{c}{4f_{dh}} \qquad (7.16)$$

For example, if the highest Doppler frequency encountered is 5 kHz, the maximum range of view is 7.5 cm. Problem 7.9 gives further numerical examples of the tradeoffs in the design of a pulsed Doppler instrument.

Detection Circuitry and Range Gate Electronics

Since Doppler frequency shifting is equivalent to phase modulation of the scattered signal, the measurement of this shift can be accomplished by using a phase detector circuit which compares the incoming signal with the master oscillator to produce a voltage output proportional to the phase angle between the two. This phase detection is most successful if the phase of the received echo is well defined. This is the case when the echo has been reflected from a limited volume of scatterers that all possess approximately the same velocity—a requirement met by the focused-beam pulsed Doppler device with its good axial and lateral resolution. Such circuits are not as useful in cw Doppler instruments where the returned signal is a combination of echoes from both moving and stationary objects near and far away, and whose overall phase angle is therefore little affected by the Doppler shift from a small component of moving scatterers.

Because the phase detector circuit is only sensitive to phase differences, the amplitude variation of the received signal is not of importance. Therefore, the received waveforms are usually greatly preamplified and then limited or clipped at some level prior to phase detection. This "hard limiting" improves the accuracy of the phase determination by removing amplitude modulation in the Doppler return.

When directional information is desired for the flow, a quadrature phase detection scheme (similar to the concept used for cw Doppler) may be employed to discriminate between reverse and forward velocities.

The output from the phase detection stage is passed to the range-gated receiver. Here, a time-windowed sample of the Doppler signal is taken; the time delay between pulse emission and the sampling gate is proportional to the depth of propagation. As mentioned earlier, several samples are needed to build up the Doppler frequency curve. This is accomplished, for example, by a sample-and-hold circuit, followed by a filter to remove the sampling frequency component.

Coverage of the entire range may be done in one of two ways: Either a single window is slowly swept in delay time, or multiple fixed-delay gates are used to collect the data in parallel. The first method is less complex in that it requires only one gate. But since several samples must be collected at each depth, the time required to scan the entire propagation range is rather long, spanning many heart cycles. Beat-to-beat velocity variations will then show up as extra irregularities in the velocity profile. Multiple range gates (perhaps 16) allow collection of complete profiles within one heart cycle.

The lower cutoff frequency of the filters in the signal processing stage should be low enough to pass the lowest important Doppler shift frequency, but high enough that nonmoving scattering components are removed and velocity variations making up the heart cycle are followed and not damped out. Typically, a lower cutoff frequency of 50 Hz will be satisfactory.

Spatial Resolution

The spatial resolution cell of the pulsed Doppler instrument is fixed by the same considerations that apply to a conventional pulsed imager. In the lateral direction, the beam size and corresponding lateral resolution may be improved by focusing the transducer. The axial resolution is determined by the duration of the transmitted burst, which in turn is dependent on the length of electrical excitation and Q of the transducer.

The width of the sampling gate in time should be shorter than the transmitted acoustic pulse to avoid degrading the axial resolution, on the order of 0.5 μs or less. A typical resolution volume is 2 mm long by 3 mm^2 in cross section.

Uses of Pulsed Doppler

Identifying the location of the Doppler return adds significant diagnostic value to this velocity instrument. It is possible to delineate the boundaries of flowing streams and to distinguish flow in neighboring vessels. By scanning the beam along the direction of a vessel's axis, the angle of inclination with respect to the skin's surface can be estimated; this angle may then be used in Equations (7.10) and (7.13).

One very important application of pulsed Doppler data is to estimate the *volume* flowrate, not merely velocity. Physiologically, volume flowrate is a valuable indicator of a vessel's ability to deliver oxygen to tissue regions. Instantaneous volume flowrate Q is given by

$$Q = \int_A V \, dA = \bar{V} A \qquad (7.17)$$

where $\bar{V}$ is defined as the spatially averaged velocity over the cross section A of the vessel at each instant of time. To accurately determine Q, the velocity profile throughout the vessel's cross section must be found as well as the area A. With care, cw Doppler flowmeters can give an estimate of $\bar{V}$, but they do not provide information on vessel size. Pulsed Doppler gives profiles along one line through the vessel and therefore can also measure diameter; with the assumption of circularly uniform flow, a good estimate of A, $\bar{V}$, and Q is obtained (see Problem 7.10).

Velocity profiles also reveal flow abnormalities coincident with plaque deposits, stenosis, or aneurism in vessels.

7.4.3 Duplex Scanners

One of the most sophisticated ultrasonic bioinstruments is the duplex scanner. By combining pulsed Doppler with conventional B-scanning, a powerful dual-modality instrument capable of superimposing a great deal of acoustic information about the body is obtained. The pulsed Doppler produces information on velocity versus depth, while the B-scan gives a two-dimensional map of the reflected amplitudes from tissue interfaces and impedance boundaries (such as vessel walls).

The B-scanning can be accomplished by any of the means discussed in Chapter 6 on real-time imaging: mechanical motion, linear arrays, or electronic phased arrays. Each approach has its own advantages and disadvantages. For example, a single transducer capable of both B-scanning and pulsed Doppler can be mechanically rotated to produce conventional B-sector pictures, then can be stopped to give pulsed Doppler along any one chosen direction. A variation of this scheme is pictured in Figure 7.12a, which shows a head with separate simultaneously excited transducers, one for mechanical B-scanning and one for pulsed Doppler.

If a linear array is used to produce the B-scan, a separate transducer mounted alongside the array is also usually needed for the Doppler beam; otherwise, each element of the array would have to possess individual

7.4 DOPPLER IMAGERS

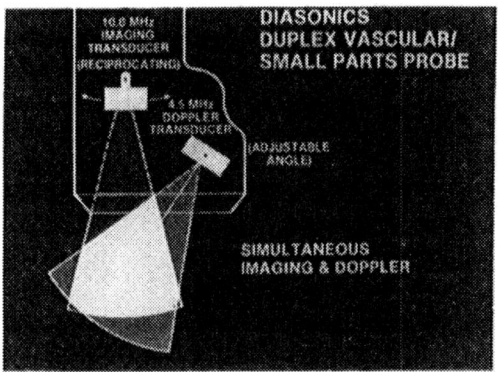

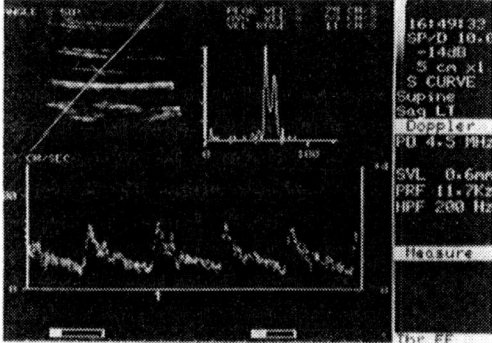

Figure 7.12 (a) A duplex probe for simultaneous B-mode mechanical scanning and pulse Doppler of peripheral vessels using separate transducers. (b) A duplex display showing the B-mode image in the upper-left of the picture and a pulsed Doppler spectrum vs. time in the lower portion. The point of Doppler measurement is indicated by the crossed lines in the B-mode image. Photographs courtesy of Diasonics, Inc.

Doppler capability, making the system very complex. Theoretically, at least, the phased array technique would be the most flexible, interlacing pulsed Doppler along any electronically chosen line with a complete sector B-scan (as is done now with M-mode lines within the sector). This promise may be difficult to achieve, however, due to phase noise in the delay modules associated with each channel.

The display can take two forms: The Doppler graph of velocity spectrum or average velocity at a selected point may be presented alongside the conventional B-sector, as in the example shown in Figure 7.12b, or the two maps may be overlaid on top of one another. In the second case the Doppler information can be encoded as color changes or as varying lengths of some pictorial element such as small arrows on the screen.

As an example of the clinical value of a duplex image, consider the case of obstructing plaque that is deposited in diseased carotid arteries and that occurs in various stages of hardness, from soft, fatty deposits to harder, calcified regions. The softer the plaque, the more difficult it is to distinguish from normal vessel wall and blood in a conventional B-scan. But when combined with views of the restricted flow as provided by the pulsed Doppler portion of a duplex image, the diagnosis of the degree of stenosis is often more accurate than with either modality by itself.

Catheter-Tip Doppler

In a broad sense, the idea of placing a Doppler-sensing transducer in combination with a diameter-measuring array at the tip of a catheter falls into the classification of duplex imaging. Figure 7.13 shows the scheme. The catheter is introduced into a vessel (the pulmonary artery, for example) where it positions itself somewhere within the vessel's lumen. A Doppler transducer radiating longitudinally in the direction of flow measures the velocity of the blood. Depending upon the beam divergence, vessel diameter, position of the tip, and flow profile, this velocity is more or less a measure of the cross-sectional average velocity $\bar{V}$ in the vessel.

A second set of transducers arranged in a ring pattern around the

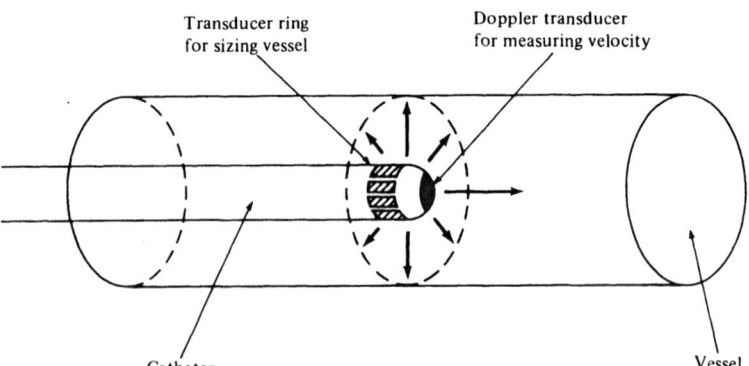

Figure 7.13 Shown is a catheter tip technique which combines a Doppler transducer with a ring array for gauging vessel area. Together the two measurements provide the ingredients necessary to determine volumetric flowrate Q.

circumference of the catheter is used in the pulse echo mode. Each element is excited sequentially, and the echo arrival times indicate the distances to the vessel wall. From these, the area A can be determined, and the product $A\bar{V}$ gives the volumetric flowrate Q. When placed in the pulmonary artery, this device will therefore measure cardiac output.

It should be noted that if the catheter axis is not aligned precisely parallel to the flow direction, the Doppler-derived velocity will underestimate the true velocity due to the angle term in the Doppler equation. However, the area measured will be overestimated by the same factor, and therefore the product, which gives flowrate, will be unchanged by tilt (see Problem 7.11).

A drawback to this device is the large number of transducers and electrical connections required in the catheter. Competing techniques for measuring cardiac output in the pulmonary artery include dye dilution and thermodilution catheters. Both of these techniques use less complex catheter designs but will not follow the rapid variations of flow within the cardiac cycle that the Doppler device does.

7.5 TRANSIT TIME FLOWMETER

We now investigate techniques for measuring flow that do not rely upon the Doppler principle. One very straightforward configuration for detecting flowrates is based upon the fact that the velocity of acoustic waves propagating in a medium is a constant c when referenced to a frame that is stationary with respect to the average position of the fluid particles. But if the fluid, in turn, possesses a velocity V with respect to an outside frame such as the walls of a vessel, then the net velocity of the waves with respect to the outside frame will be a vector summation of the two velocities. Consider the situation shown in Figure 7.14, where a pulse of ultrasound is transmitted from transducer A to transducer B positioned obliquely across the vessel. The pulse propagates a distance W in stationary tissue with phase velocity c and travels a distance L through the moving fluid. Due to the relative motion of the fluid, the effective phase velocity of the waves in the fluid with respect to the transducers will be

$$c_{\text{eff}} = c + V\cos\theta \tag{7.18}$$

where $V\cos\theta$ is the component of V along the line of propagation from A to B. Since the transit time τ_{AB} for the pulse from A to B is the distance between the transducers divided by the effective phase velocity,

$$\tau_{AB} = \frac{L}{c + V\cos\theta} + \frac{W}{c} \tag{7.19}$$

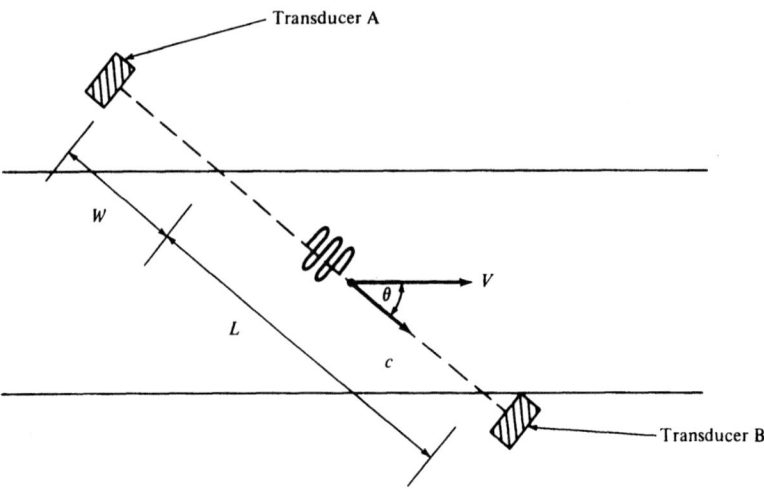

Figure 7.14 Transit time ultrasonic flowmeter, where the difference between upstream and downstream pulse propagation times is related to fluid velocity V.

Any variation in V is therefore seen as in a change in the transit time. Unfortunately, the time change is not linearly related to the fluid velocity because of the position of the V term in the denominator of Equation (7.19), and since usually $c \gg V$, the change in τ_{AB} is very small and difficult to detect. A more useful relationship may be derived by noting that the upstream transit time from B to A is given by

$$\tau_{BA} = \frac{L}{c - V\cos\theta} + \frac{W}{c} \qquad (7.20)$$

The *difference* between upstream and downstream propagation times is therefore

$$\Delta\tau = \tau_{BA} - \tau_{AB} = \frac{L}{c - V\cos\theta} - \frac{L}{c + V\cos\theta} + \frac{W}{c} - \frac{W}{c}$$

$$= \frac{2LV\cos\theta}{c^2 - V^2\cos^2\theta}$$

so

$$\boxed{\Delta\tau \approx \frac{2LV\cos\theta}{c^2}} \qquad (7.21)$$

where the last approximation is valid for the case of $c \gg V$.

7.5 TRANSIT TIME FLOWMETER

Now the difference in propagation times is proportional to the fluid velocity. Note that the propagation delay through the stationary medium cancels out when the difference is taken, so W does not appear in the final result. The other terms in the equation (L, $\cos \theta$, and c) are constants of a particular flowmeter configuration and are accounted for by calibration of the instrument. An exception to this last statement is the use of the transit time flowmeter in respiratory gas measurements, since in that application the composition and humidity of the gases flowing in and out of the lungs may change during different therapeutic actions when the patient is attached to artificial ventilation. For example the composition of the inspired air may be changed from moist room air to pure dry oxygen with a corresponding change in c in the gas. This will alter the calibration of the instrument, unless some compensation is made.

The actual magnitude of $\Delta \tau$ resulting from Equation (7.21) is quite small in most applications, being in the range of a few nanoseconds for blood flow (see Problem 7.12). Thus, the stability of the electronic circuitry detecting the arrival times of each of the pulses must be good, and the signal-to-noise ratio of the received pulses must be high enough to allow an accurate determination of when the pulse has been received. To increase accuracy, many transit times may be averaged together; the pulse repetition rate can be fast enough (several kilohertz; see Problem 7.12) that hundreds of pulses may be averaged within a reasonable time constant for the instrument. For random noise fluctuations in the measured transit times of individual pulses, the accuracy of the averaged result increases proportionally to the square root of the number of pulses taken in the average.

The derivation leading to Equation (7.21) assumed a uniform velocity profile within the vessel. When the flow is not uniform (as it seldom is), the velocity V in Equation (7.21) must be interpreted as the velocity averaged over the path of propagation of the beam across the vessel. If the beam diameter is much narrower than the diameter of the vessel, this velocity is a *line* average along the path of the beam from one side of the vessel to the other. The line average of the velocity is not equal to the cross-sectional average $\bar{V}$ which appeared in Eq. (7.17); $\bar{V}$ is the velocity averaged over the entire two-dimensional cross section of the flow, not just along one line. For the case of laminar flow, where the velocity profile is parabolic, the line average velocity V in Equation (7.21) is equal to $1.33\bar{V}$; this result is derived in Problem 7.13.

A major disadvantage of the transit time flowmeter for human use is the need to access both sides of the vessel for placement of the transducers, usually requiring a cutdown to expose the vessel. It has its primary application in animal studies (both acute and chronic implants) and invasive measurements. In these situations it competes with the Doppler ultrasound flowmeter and electromagnetic cuff flowmeter. Transit time devices have the advantage that they measure transmitted ultrasound, not scattered power, and therefore the signal strength is much higher. Inasmuch as the path delay through any stationary medium is cancelled when the transit time difference is taken, the transducers do not have to be in close contact with the vessel walls, as is the case with electromagnetic flowmeters. Therefore, the vessel is unrestricted, as long as acoustic contact by some coupling medium is maintained. Also, the transit time flowmeter can discriminate bidirectional flow and does not suffer from zero drift as electromagnetic sensors sometimes do.

One current embodiment of the transit time flowmeter (shown in Figure 7.15) takes the concept one step further. It transmits a *wide* beam of ultrasound through the vessel, wide enough to cover the entire diameter of the flow. The velocity measured is therefore a more accurate representation of the cross-sectional average $\bar{V}$. In fact, with proper calibration, this flowmeter will read volumetric flowrates directly, independently of the size of the vessel inside the sensor. The head is loosely fitting and uses a reflector on the opposite side of the vessel to reflect the acoustic pulse from one transducer back to a second transducer on the same side. By alternating the roles of transmitter and receiver, both upstream and downstream transit times are obtained by a phase detector circuit. Subtraction gives the transit time difference, proportional to volume flowrate. Long-term stability ($\pm 2\%$) and absolute volume flowrate accuracy ($\pm 10\%$) are very good for this device.

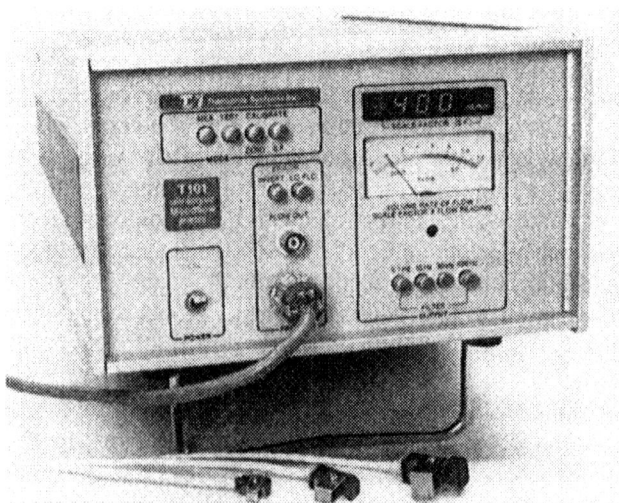

Figure 7.15 A wide-beam transit time flowmeter for invasive use which reads volumetric flowrates directly. Photograph courtesy of Transonic Systems, Inc.

7.6 VORTEX FLOWMETER

A non-Doppler technique currently being used for measuring respiratory gas flowrates relies upon the production of local eddies (vortices) in the flow of the gas immediately behind an obstruction purposely placed in the flow to cause it to be turbulent. Figure 7.16 shows a schematic of the system. Triangular struts are placed in the flow to produce vortices in the gas as it passes around a strut. The vortices are produced in pairs, and the number of pairs produced per unit time is proportional to the flowrate of the gas. This number is measured by the modulation each vortex produces in a cw ultrasound beam that is transmitted across the flow by two transducers downstream from the obstruction. The frequency of the amplitude modulation of the beam is detected by receiver electronics.

A key characteristic of the vortex shedding phenomenon is that the frequency of vortex production is independent of the fluid properties in the flow. Thus, when used as a respiration monitor, this flowmeter will read accurately even though the gas composition to the patient is changed from time to time. The axial spacing between each vortex is always approximately 2.5 times the width of the strut perpendicular to the flow direction.

Combined with digital electronics, the vortex flowmeter can provide many of the measurements important to respiratory function monitoring. These include tidal volume (the gas volume inhaled or exhaled per breath), the minute volume (the total volume inspired during a period of 1 minute),

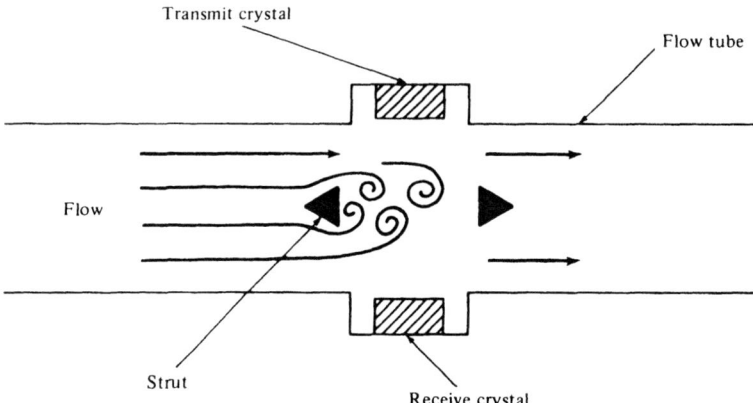

Figure 7.16 Air flow past the struts inside the vortex-shedding flowmeter will set up vortices in the region behind the strut at a rate proportional to flow velocity. These vortices interrupt the amplitude of an ultrasonic beam passing across this region at a frequency proportional to velocity.

and the respiratory rate (in breaths per minute). Due to the symmetric nature of the device, it is independent of flow direction. Also, since only the frequency of interruption of the beam is detected, not its absolute amplitude, this sensor is unaffected by amplitude drifts and by fouling of the transducers by moisture or other contamination in the exhaled air.

PROBLEMS

7.1. (a) The cardiac output (volume of blood pumped by the heart per minute) of a typical female adult is about 4.5 liters/minute. Estimate the peak blood velocity V_p in the aorta (the heart's output vessel) by assuming a uniform flow profile across the aorta, a typical aorta diameter of 2 cm, and time variation of blood velocity modeled by the simple triangular wave shown below:

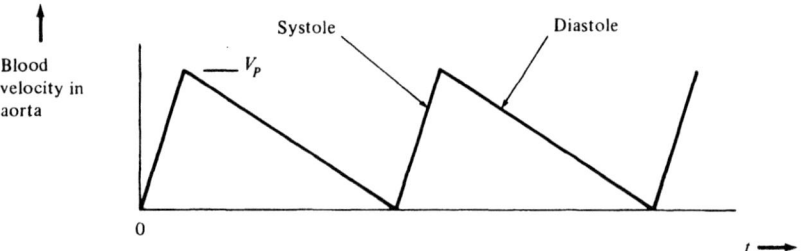

(b) Compare V_p to c. Is $V \ll c$?
(c) Show that Equation (7.8) follows from Equation (7.7) for $V \ll c$.

7.2. For the situation where the moving scatterer is embedded in a moving fluid and both are moving at velocity V with respect to the source and receiver, find the Doppler shift f_d using the same angles as shown in Figure 7.1. Simplify your answer by using the approximation $V \ll c$ and compare to Equation (7.8). This exercise applies to red blood cell scattering in a flowing blood stream.

7.3. The transcutaneous ultrasonic blood flowmeter shown below is used to measure blood velocity in a shallow artery. Assume the blood velocity in the vessel is uniform throughout the cross-sectional area of the artery (in most cases a poor assumption) and that its magnitude as a function of time is as shown. Plot the value of the Doppler shift frequency f_d as a function of time if the source frequency is $f = 4.0$ MHz.

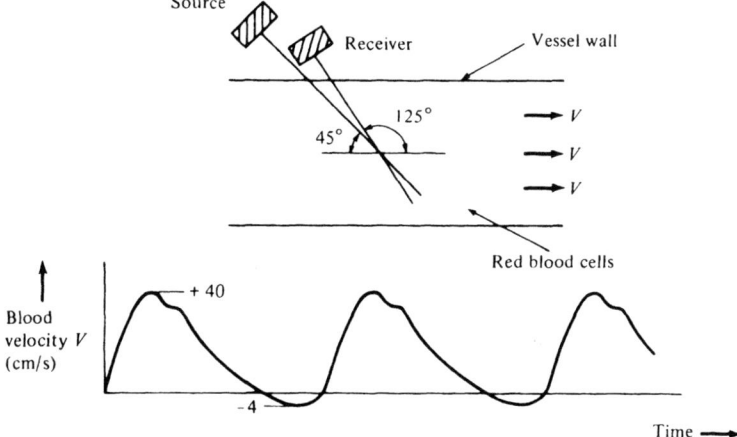

7.4. Find the percentage error in measured blood velocity for each of the following two Doppler flowmeters if they are mispositioned ±10° away from the angle for which they were calibrated (for simplicity, let the transmit and receive transducers be the same element):
  (a) A hand-held peripheral vascular flowmeter calibrated assuming an angle of 45°.
  (b) An aortic Doppler imager calibrated for an angle of 0° (i.e., viewing along the axis of the ascending aorta).

7.5. Estimate the extent of spectral broadening in the frequency return of a cw Doppler flowmeter due to the following two factors (give the width of broadening for each factor as a percentage of the center frequency and assume that the transmit and receive functions are performed by the same transducer element):
  (a) The incident and receive beams are not composed of parallel rays but rather are focused to a spot 4.0 cm away from the transducer. The transducer diameter is 1.0 cm, and its axis forms an angle of 60° with respect to the axis of the blood vessel.
  (b) The effective viewing volume has a dimension of 2.5 mm along the direction of blood flow, so each red blood cell is viewed only for a finite amount of time. The incident frequency is 3 MHz.

7.6. This problem analyzes the mixing scheme for the cw Doppler instrument shown in Figure 7.5. Assume for simplicity that there is only a single Doppler-shifted frequency component present. The receiver signal input voltage to the mixer is given by (neglecting constant phase factors)

$$V_1 = A_1 \cos 2\pi f_i t + A_2 \cos(2\pi f_i - 2\pi f_d)t$$

where A_1 is the signal amplitude reflected from stationary objects and A_2 is from the uniformly moving scatterers. The transmitter oscillator's input voltage to the diode mixer is

$$V_2 = A_3 \cos 2\pi f_i t$$

The mixer stage applies the sum of the two input voltages $V_t = V_1 + V_2$ to the diode, whose output current can be modeled as a nonlinear square law device:

$$I = KV_t^2$$

(a) Show that the diode output current I contains components whose frequencies are all the possible sum and differences of the input frequencies. [*Hint:* Use the trigonometric identity $2 \cos A \cos B = \cos(A + B) + \cos(A - B)$]. Find the amplitudes of each of these components, and sketch a spectrum of the current showing the frequency position of each of the components. Translate all components to positive frequencies.

(b) If the Doppler shift is positive $(+f_d)$ instead of negative $(-f_d)$, what changes would there be in the above results?

7.7. (a) In most intermediate-size arteries in the body, the peak blood velocity during systole is between 2 and 100 cm/s. Estimate the range of the Doppler shift frequency produced by a cw ultrasonic flowmeter. Are these frequencies in the audible range?

(b) A certain artery 5 mm in diameter expands in cross-sectional area by 20% during a 50-ms segment in peak systole. Estimate the transverse velocity of the arterial wall and estimate the Doppler shift corresponding to this wall motion. Does it overlap in frequency with the Doppler shift due to blood flow?

7.8. Calculate the optimum incident frequency for a Doppler blood flow instrument by finding the frequency at which the intensity of the received signal is a maximum for a given depth of the scattering site. Assume that the red blood cells within the scattering volume act as Rayleigh scatterers (f^4 dependence) and that the attenuation of the intervening tissue has a linear frequency dependence of 1 dB/cm per MHz for a one-way path. Find the optimum frequency for two cases:

(a) Shallow arteries lying at a depth of 1 cm.
(b) Deeper vessels at a depth of 5 cm.

7.9. Design a pulsed Doppler blood flow instrument for the following specifications:

> Axial resolution = 1 mm
> Angle between beam and flow = 60°
> Maximum velocity = 150 cm/s
> Viewing range = 5 cm
> Lower cutoff frequency of output bandpass filter = 100 Hz

For your design, specify the incident ultrasound frequency, the number of cycles in each incident pulse, the pulse repetition rate, and the minimum velocity this instrument will measure.

7.10. The pulsed Doppler flowmeter below gives the velocity profile shown at one instant of time. Estimate A, Q, and $\bar{V}$ for the vessel imaged. *Hint:* Assume parabolic flow of the form

$$V(r) = V_{max}\left(1 - \frac{r^2}{a^2}\right)$$

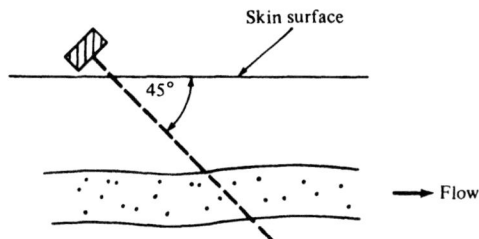

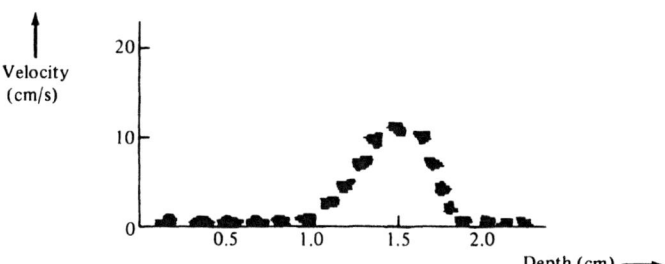

7.11. Show mathematically that the amount of angular misalignment between the axis of the catheter tip flowmeter and the vessel in which it is inserted will not affect its measurement of volumetric flowrate.

7.12. Consider the transit time flowmeter configuration shown below for measuring blood flow in an artery:

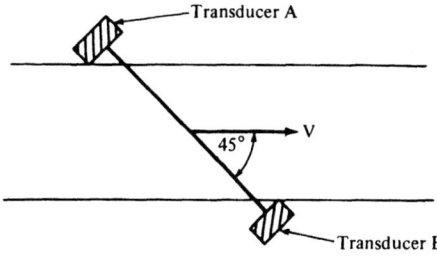

Assume that the flow profile is uniform. The diameter of the vessel is 1.0 cm.
(a) What is the time difference $\Delta\tau$ between upstream and downstream pulse transit times if the volumetric flowrate in the vessel is 0.5 liter per minute?

(b) Estimate the maximum pulse repetition rate that could be reasonably used in the above system such that successive pulses (and their multiple-reflected echoes) would not overlap. List your assumptions.

7.13. Show that if a transit time flowmeter is used to measure a vessel with a parabolic velocity profile (laminar flow), then V in Equation (7.21) must be replaced with $1.33\bar{V}$, where $\bar{V}$ is the velocity averaged over the cross-sectional area of the flow. (*Hint:* See Problem 7.10.)

7.14. Calculate the frequency of the modulation of the ultrasound beam in a vortex flowmeter if the gas flowrate is 18 liters per minute, the flow tube cross-sectional area is 2 cm², and the triangular strut measures 2 mm on a side. Assume that the flow profile is uniform across the tube.

Chapter 8

The Safety and Measurement of Ultrasound

8.1 INTRODUCTION

One of the well-promoted advantages of ultrasound in medical imaging is its apparent safety. This feature has led to its extensive replacement of X-radiation for obstetrical scanning, for example. But ultrasound at high power levels does in fact inflict considerable tissue damage and has been purposely used at high power levels for tissue destruction and ablation. Thus, the question presents itself: What is the relationship between the incident power level (actually power density is probably a more relevant quantity here) and tissue hazard? Also, what effects do frequency, pulse length, and exposure time have on tissue toxicity? The ultimate answers to these questions are still being sought in numerous investigations such as in vitro cell exposures, animal studies, and long-term epidemiological studies. It will undoubtedly be a long time before final definition is found, but at present the evidence indicates that at the low exposure levels employed for diagnostic imaging, ultrasonic hazards are minimal and are far outweighed by the potential benefits derived from the image data obtained.

In this chapter we investigate some of the possible mechanisms by which ultrasound exposure may cause changes in tissue, discuss ways of measuring ultrasonic power relevant to biological effects, and present guidelines which have been suggested for determining the domain of safe ultrasonic exposure.

8.2 POSSIBLE MECHANISMS OF DAMAGE

In liquid environments the passage of ultrasound produces several effects which may have harmful results. These include both thermal and nonthermal action on the liquid and its contents. It is still unclear to what degree these mechanisms are present in the cellularly organized structure of tissue, but they represent possible sources of damage.

Absorption

The loss of energy from the propagating wave to the molecules of the supporting medium may be partitioned into two ultimate results: (1) a reversible increase in the vibrational and rotational energy of the molecules, sensed as a temperature rise of the material; and (2) permanent modification of the molecular structure such as broken bonds. The first action is classified as *thermal*, and the temperature increase may produce subsequent damage such as protein modification if the increase is large enough. Tissue has relatively high absorption and low thermal conductivity, which accentuate the thermal rise. The second action, direct molecular change, is a *non-thermal* effect.

In living tissue it appears that large protein molecules are responsible for the majority of absorption via various relaxation processes. It is comforting to note, however, that the phonon energy in ultrasound is insufficient to cause dangerous ionizing damage (that is, ejecting electrons from their bound orbits around atoms) as occurs with X-radiation.

Mechanical Action

The vibration and pressure associated with the ultrasound wave may have a direct effect on the structure of cells and tissue. For example, cell walls and internal organelles may be disrupted by the passage of high-intensity ultrasound through shearing and streaming forces. This damage is nonthermal in nature.

Chemical Effects

The chemistry of substances may be affected (i.e., the chemical dynamics and pathways may be changed) by the pressure and temperature changes attendant with ultrasound.

Cavitation

In liquids the phenomenon of cavitation plays a major role in disrupting the materials suspended or dissolved in the liquid. Cavitation occurs when small, submicroscopic bubbles of dissolved gas in the liquid coalesce to form larger bubbles which then either may stabilize in size or may catastrophically explode or collapse. Even if the bubble is stable, when it is at a size that is resonant with the incident wavelength, large amplitude motion will be present in the neighborhood of the bubble, leading to potentially severe damage to molecular configurations and other structures.

It is unclear whether cavitation takes place to any extent in the contained liquid portions of the compartmentalized structure characteristic of living tissue. Also, it takes a finite time for cavitation to be produced, so the short pulses of ultrasound used in echo imaging (typically a few microseconds) may not be sufficiently long to set up cavitation effects. Thus, the exact role of cavitation in tissues is still unknown.

8.3 MEASURING ULTRASOUND EXPOSURE LEVELS

This section discusses various ways to measure the exposure of tissue to ultrasonic power. The power pattern of the irradiating beam generally varies both as a function of time and space. With regard to each of these two variables, either the instantaneous (peak) value or an appropriate average over the variable may be given, leading to a possibility of four different combinations for characterizing the power. Table 8.1 delineates these four combinations. Which of the possible formulations most accurately describes the exposure will depend upon whether thermal or nonthermal effects are being considered, since the time constant for the onset of these two categories is much different.

TABLE 8.1 THE FOUR POSSIBILITIES FOR SPECIFYING POWER EXPOSURE LEVELS

TIME	SPACE	
	Peak	Average
Peak	SPTP	SATP
Average	SPTA	SATA

Thermal Exposures

In thermal exposure calculations the time constants are relatively long due to the thermal mass of the tissue; it may take a minute or so to reach a damaging temperature. During this time, the temperature produced by any localized power deposition peaks will be spatially spread out via thermal diffusion and conduction. It is therefore reasonable that the formulation that most accurately describes thermal exposure is the incident beam power density time-averaged over a moderately long time period and, in addition, space-averaged over a cross-sectional area which extends just far enough to include thermal spreading (SATA).

Figure 8.1 shows separately the temporal and spatial variations typical of an incident ultrasound beam used in diagnostic imaging. The time average of power density I_{ta} relevant to thermal exposure is given by

$$I_{ta} = \frac{1}{T_1} \int_0^{T_1} I(t)dt \tag{8.1}$$

where T_1 is of the order of the thermal time constant, generally many pulse periods long. When applied to a repetitive pulse of duration t' and period T similar to that shown in Figure 8.1, this equation becomes

$$I_{ta} = \frac{1}{T} \int_0^T I(t)dt \approx \frac{I_{peak}}{2} \frac{t'}{T} \tag{8.2}$$

The factor of 2 accounts for the time average over each ultrasound oscillation cycle (see Figure 3.1). The ratio t'/T is known as the *duty cycle* of the pulse train and gives the proportion of "on" time of the pulse. Since diagnostic imaging uses very short pulse lengths to improve axial resolution, the duty cycle is very low, and thermal effects are generally small.

The spatial average of power density is given by an integral over a plane perpendicular to the direction of propagation of the beam. For a z-directed beam whose pattern is approximately circular in shape,

$$I_{sa} = \frac{1}{\pi R^2} \int_0^R \int_0^{2\pi} I(r, \theta) r \, dr \, d\theta \tag{8.3}$$

where R is the radius of the area over which a localized increase in temperature spreads by thermal diffusion and conduction during the time T_1.

As mentioned earlier, for thermal exposure calculations the combined temporal and spatial average is generally needed. In this case the averaging indicated in Equation (8.3) should follow the averaging of Equation (8.2); that is, the result of Equation (8.2) as a function of space, $I_{ta}(r, \theta)$, is placed inside the integral of Equation (8.3). Problem 8.1 is an exercise in the calculation of a combined temporal and spatial average of power density using values typical for medical ultrasonic imaging.

8.3 MEASURING ULTRASOUND EXPOSURE LEVELS

(a)

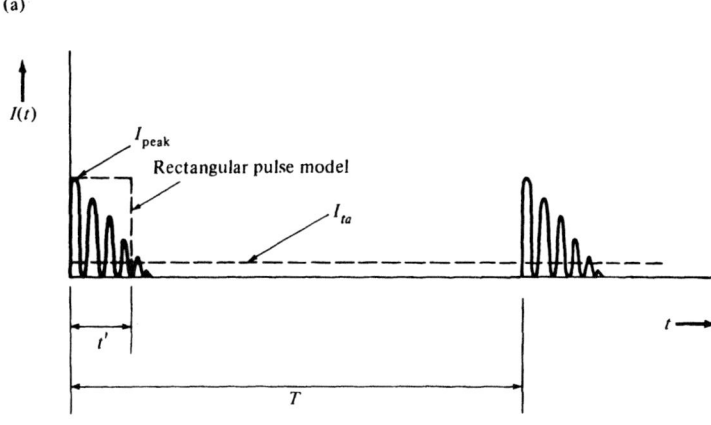

(b)

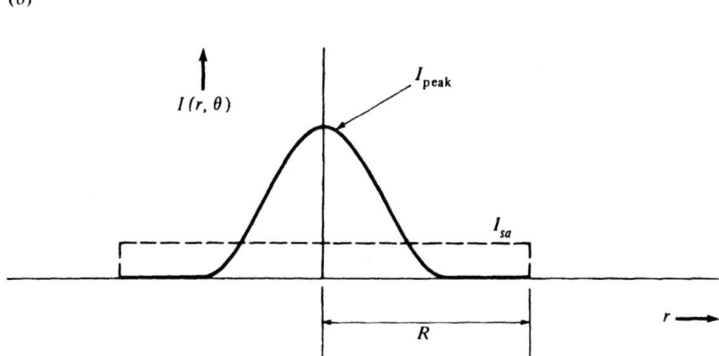

Figure 8.1 (a) A time trace typical for ultrasound pulsed exposure. I_{ta} is the power density time-averaged over several pulse periods T. (b) A spatial pattern typical of an ultrasound beam. I_{sa} is the spatially averaged power density averaged over an area of radius R.

NonThermal Exposures

As opposed to thermal hazard effects, for nonthermal effects the time constants are very short—essentially instantaneous because mechanical damage and molecular deformation can happen on a time scale comparable to an ultrasonic oscillation period. Therefore, the intensity value which is important in predicting nonthermal damage is the *peak* power density, both temporal and spatial (SPTP). Any safety standard that includes possible nonthermal as well as thermal effects should account for the peak power density in the beam, not just the average power density. Current hazards research is aimed at better defining such a standard.

8.3.1 Measurement Techniques

The intensity of a machine's beam may be measured in the laboratory or clinic by three general techniques. Each method possesses a different capability for measuring either peak or average power density of the beam.

Hydrophone

A separate small transducer called a hydrophone may be systematically moved about in the field of the transducer under test while its voltage output, proportional to received pressure, is monitored on an oscilloscope. The hydrophone, being very small in diameter, approximates a point receiver for accurately mapping the spatial pattern. The small diameter also produces a broad angular cone of reception for the hydrophone, rendering it insensitive to slight angular misalignment. The resonant frequency of the sensing transducer is designed to be much higher than the range of test frequencies encountered, so the hydrophone's resonant frequency response does not appreciably distort the received signals.

Figure 8.2a shows the general technique of scanning a beam pattern using a hydrophone in a water tank. Both the spatial and the temporal patterns of the beam may be obtained, yielding peak values for each during the scan. Figure 8.2b shows an experimentally derived spatial pressure pattern from a large, square transducer as an example of the method.

Although the hydrophone possesses good peak spatial and temporal resolution, its absolute accuracy is sometimes questionable. When new, it may be calibrated against a traceable standard, but drifting with time and temperature will change the calibration. Terminating it in different amplifiers may also affect calibration. Also, the hydrophone's point sensitivity may work against it by giving too much susceptibility to localized variations caused by coherent interference or "speckle".

Absorbing Temperature Probe

This technique uses a small temperature sensor, such as a thermocouple or thermistor, coated with an absorbing material. The amount of incident power absorbed by the coating results in a proportional temperature rise which is measured by the sensor. It is similar to the hydrophone technique

8.3 MEASURING ULTRASOUND EXPOSURE LEVELS

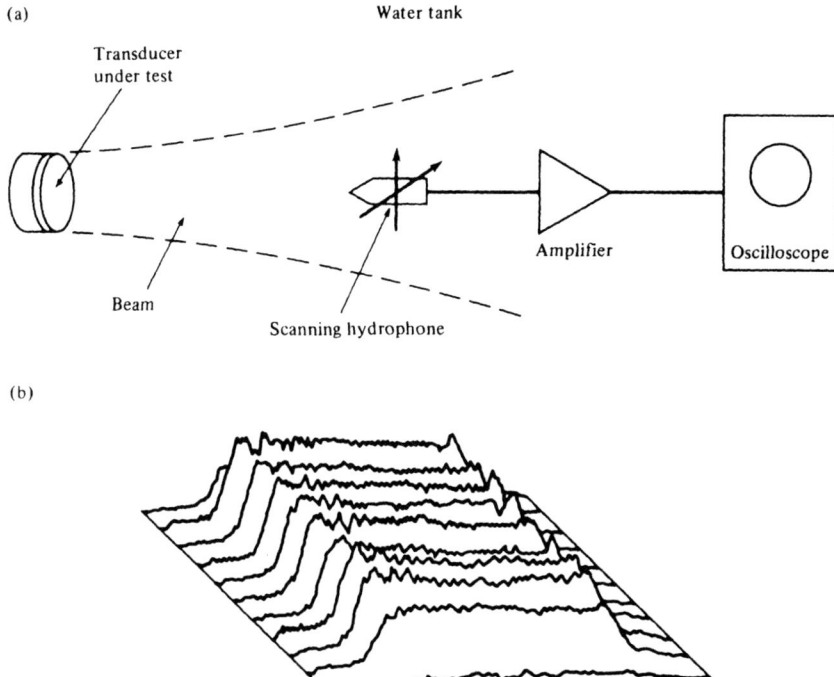

Figure 8.2 (a) A hydrophone may be used to scan the beam of a test transducer, yielding both time and space patterns. (b) An example of the spatial pressure pattern experimentally determined by hydrophone scanning a 3-MHz, 10 cm × 10 cm square transducer at a distance of 18 cm. From P. C. Peterson and D. A. Christensen, *Acoustical Holography* 6 (1975), 711–739, Plenum Press.

inasmuch as it is a point measurement and scanning is accomplished as in Figure 8.2. However, it measures power directly, not pressure, and its time response is much slower than that of the hydrophone. It therefore will not follow the peak values of short pulses but instead gives time-averaged readings. Because of variations in the coating material, these probes also need to be calibrated against a trusted standard.

Radiation Force

The third method differs from the first two in that it is an absolute technique which needs no ultrasonic calibration. It relies upon the principle that an acoustic beam carries an average momentum per unit time past a plane perpendicular to its direction of propagation:

$$\text{Momentum} = \frac{I_{\text{ave}} A}{c} \tag{8.4}$$

where I_{ave} is the beam's average power density and A is its cross-sectional area. (Although this equation is strictly true only for continuous plane waves in a constant-compressibility nonviscous fluid, it holds with good accuracy in water for the beams generally encountered in medical imaging.)

If the beam is reflected by a target, as indicated in Figure 8.3a, the momentum direction will be altered, causing a force $F = \Delta$ momentum/Δt on the target which can be measured by a microbalance. If the target is totally reflecting and is tilted at an angle θ with respect to the beam (usually done so that the reflected beam may be deflected and trapped to avoid multiple reflections), then the force normal to the target is

$$F_n = \frac{2I_{ave} A \cos\theta}{c} \tag{8.5}$$

where the factor of 2 accounts for the reflection. (If the target is completely absorbing, this factor reduces to unity.)

Figure 8.3b shows one embodiment of the radiation force measurement technique using a V-shaped target suspended from the underside of a precision microbalance. Since the normals to the two target surfaces form an angle θ with respect to the direction of force measurement, the vertical force measured by the balance is

$$F = F_n \cos\theta = \frac{2I_{ave} A \cos^2\theta}{c} \tag{8.6}$$

The microbalance has high accuracy in the measurement of the force F, and the angle θ and phase velocity c are known precisely; therefore, the average power of the beam, $I_{ave} A$, may be determined with good accuracy. The target and balance have relatively slow responses; as a result this technique will give a time, as well as a spatial, average of beam power.

Other configurations of the radiation force technique include measurement of the horizontal displacement of a suspended target (such as a ball) under the influence of gravity (see Problem 8.3) or the vertical displacements of a floating target. Commercial devices using this technique are now available for calibration of clinical imaging transducers.

8.4 SAFETY STANDARDS

The present state of knowledge regarding the absolute safety of ultrasound at the intensity levels encountered in clinical diagnostic imaging is still being refined. Enough is now known, however, to conclude that large, short-term detrimental effects are not present at these power levels. Certainly

8.4 SAFETY STANDARDS

(a)

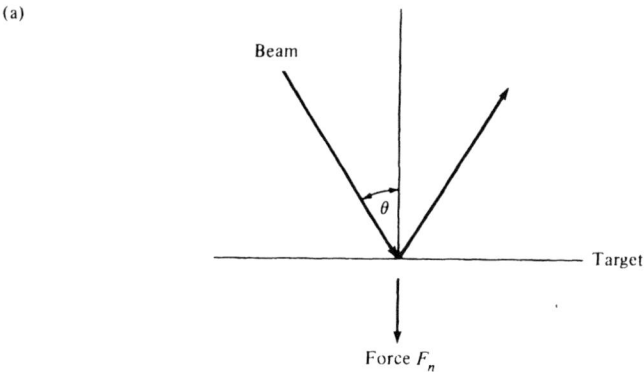

(b)

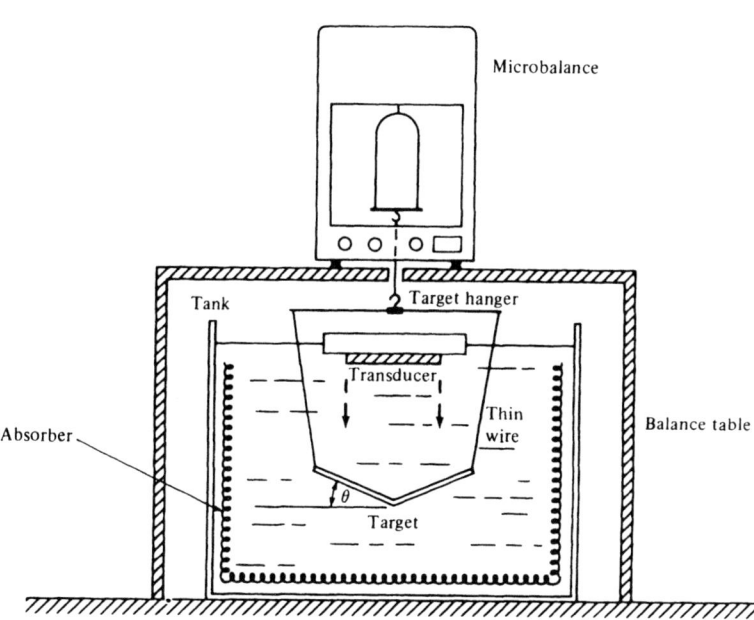

Figure 8.3 (a) The momentum change in a reflected beam will cause a radiation force on the reflector. (b) One configuration for measuring the power from an ultrasonic transducer using a target suspended from a microbalance. For accuracy, the target here contains a thin air layer to assure (almost) total reflection of the incident beam, and the reflected waves are trapped in an outer absorbing portion of the tank to avoid multiple reflections.

when there is a clinical need for the information that can be provided by an ultrasound scan, the medical benefits far outweigh the small (perhaps zero) risks involved. A 1982 panel convened by the March of Dimes concluded the following with regard to exposure of the fetus: "So far, there are apparently not even anecdotal reports of harmful effects in children exposed to ultrasound in utero."

A study by the National Center for Devices and Radiological Health of the Food and Drug Administration* recently stated: "We can be reasonably certain that acute, dramatic effects are not likely. But, studies have not been made to detect less obvious effects, and the question of subtle, long-term or cumulative effects remains unanswered. The potential for acute adverse effects has not been systematically explored, and the potential for delayed effects has been virtually ignored." Therefore, the FDA suggests that in obstetrical cases, pregnancy alone should not be considered an indication for a routine ultrasound exam.

This conservative approach is echoed by the Commission on Ultrasound of the American College of Radiologists with this advice: "Because of the controversy about ultrasound, we urge that routine ultrasound examinations of the pregnant uterus not be performed, and that use of ultrasound be limited to medically indicated cases."

It should be noted that the above cautionary advice is not based upon any comprehensive evidence of harm to humans at the power levels in question; it merely represents a better-safe-than-sorry posture until more is known.

Guidelines

At other power levels or pulsing conditions—for example, in newly developed research devices—it would be helpful to have some guidelines for whether the exposure is likely to present a danger to the patient. Unfortunately, due to a present lack of thorough studies, such a universally accepted safety standard does not yet exist. However, a formula suggested some years ago by Ulrich† has been found to be a useful guideline by several investigators. Based upon data of harmful and nonharmful effects as of 1974 using both pulsed and cw ultrasound, the line demarcating the zone of minimal hazard from more harmful exposure is shown in Figure 8.4. The "safe" region is below the line. The vertical axis plots time average intensity, regardless of whether the beam is pulsed or cw. The horizontal axis is the total exposure time to the beam at that intensity, including time between pulses, not just the pulse "on" time. Although not clearly specified,

---

* FDA Publication No. 82-8190 (Washington, D.C.: Government Printing Office, 1982).
† W. D. Ulrich, *IEEE Trans. Biomed. Eng.* 21 (Jan. 1974), 48–51.

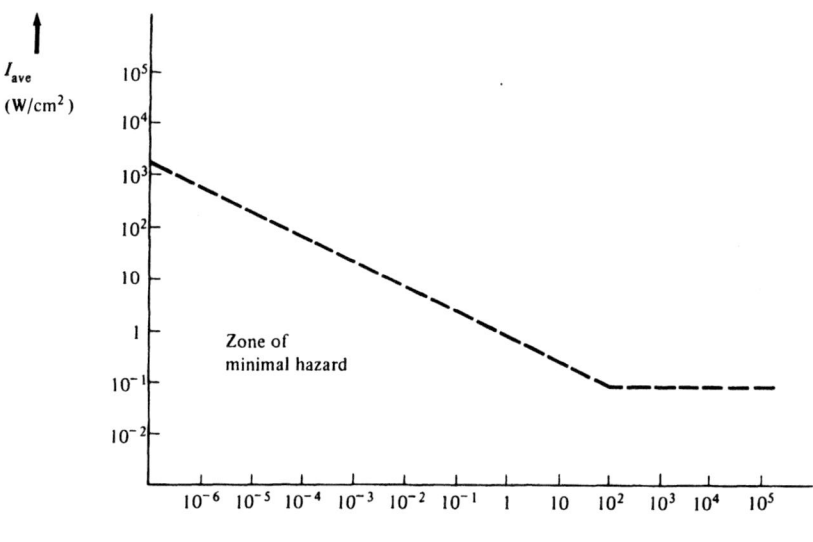

Figure 8.4 A suggested guideline for ultrasonic exposure from W. D. Ulrich, *IEEE Trans. Biomed. Eng.* 21 (January 1974), 48–51, based upon hazard studies completed at that time. More work needs to be done in the area of biological effects of ultrasound, particularly on subtle or long-term effects.

it is reasonable to interpret the intensity as a local spatial average of power density.

Figure 8.4 is meant to be applied to all regions of the body other than the eyes, and for frequencies from 0.5 to 15 MHz. Problem 8.5 shows that a typical diagnostic B-scanner falls clearly below the line, indicating probable safety. Nevertheless, much more study needs to be done to assure the safety of ultrasound with special attention paid to subtle or long-term effects.

PROBLEMS

8.1. A diagnostic imaging machine utilizes pulses of 2 MHz ultrasound that are approximately 5 pressure cycles long; the envelope may be modeled as a rectangular envelope. The pulse repetition rate is 1500 pulses/s. The spatial shape of the beam at the plane of interest is circularly symmetric and is given by

$$I(r) = I_{peak} e^{-r^2/a^2}$$

where r is the radial variable and $a = 2$ mm. The peak power density of the beam at the center of the pattern and at the point of time of maximum pulse power is

$$I_{peak} = 1 \text{ W/cm}^2$$

Also, the diameter of the circular extent of the temperature spread due to thermal diffusion and conduction is approximately 1.5 cm. Find the combined temporal and spatial average power density (SATA) for this machine.

8.2. A transducer tested by the radiation force balance method of Figure 8.3 produces a force of 20 mg on the balance. The target is a totally reflecting plane tilted at an angle of 20° with respect to horizontal. If the effective beam cross-sectional area is 7.0 cm², determine the combined temporal and spatial average power density for this transducer. According to Figure 8.4, how long can a patient be safely exposed to this beam?

8.3. The radiation force technique is used to measure the power density of a transducer by the "metal ball deflection" method shown below:

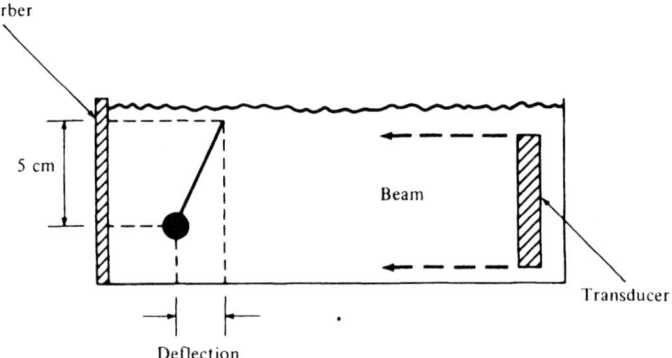

If the spherical metal ball is made of brass (diameter = 3 mm) and is deflected 0.5 mm by the radiation from the transducer, what is the power density at that point in the beam?

8.4. A certain microbalance has an accuracy of approximately 0.1 mg. When used in the radiation force method, estimate the accuracy with which the average power can be measured.

8.5. Plot on a copy of Figure 8.4 the positions of two typical ultrasound instruments:
   (a) A manual B-scanner with a peak power of 0.75 W in a beam focused to a spot 2 mm in diameter. The pulse length is 2 μs long with a pulse repetition rate of 2000 per second. Due to sweeping, the effective exposure time at any position in the tissue is about 1 second.
   (b) An ultrasound heating apparatus used for the experimental investigation of hyperthermia as a possible therapy for cancer. The total output power of the transducer is 80 W in a beam 1.5 cm in diameter. Treatment time for the cancer patient is approximately 30 minutes.

Bibliography and Suggestions for Further Reading

GENERAL

1. P. N. T. Wells, *Physical Principles of Ultrasonic Diagnosis* (New York: Academic Press, 1969). Complete coverage of topics involved in medical ultrasound.
2. P. N. T. Wells (ed.), *Ultrasonics in Clinical Diagnosis,* 2nd edition (Edinburgh: Churchill Livingstone, 1977). Good coverage of clinical applications, plus chapters on physical principles and biological effects.
3. J. L. Rose and B. B. Goldberg, *Basic Physics in Diagnostic Ultrasound* (New York: John Wiley and Sons, 1979). Covers all aspects of diagnostic ultrasound, with approximately 80 pages of laboratory experiments.
4. K. R. Erikson, F. J. Fry, and J. P. Jones, "Ultrasound in Medicine—A Review," *IEEE Trans. on Sonics and Ultrasonics* SU-21(July 1974), 144–170. Good review article on status of medical ultrasound as of 1974, including a substantial list of references.
5. L. Cromwell, F. J. Weibell, and E. A. Pfeiffer, *Biomedical Instrumentation and Measurements,* 2nd edition (Englewood Cliffs, N.J.: Prentice-Hall, 1980), 263–276. Good discussion of diagnostic considerations.
6. J. J. Carr and J. M. Brown, *Introduction to Biomedical Equipment Technology* (John Wiley and Sons, 1981), Chapter 17. Condensed but thorough discussion of physical and clinical ultrasound.
7. V. M. Albers (ed.), *Underwater Sound,* Benchmark Papers in Acoustics (Dowden, Hutchinson and Ross, 1972). A collection of early important papers in the development of ultrasound, such as attenuation, velocity, noise, scattering, transducers, and nonlinear effects.

CHAPTER 2 THE WAVE EQUATION AND ITS SOLUTIONS

1. L. A. Chernov, *Wave Propagation in a Random Medium* (New York: McGraw-Hill, 1960), Chapter 3. General derivation of the wave equation.
2. C. C. Johnson, *Biomedical Wave Effects and Instrumentation,* Department of Bioengineering, University of Utah, unpublished notes.

CHAPTER 3 IMPEDANCE, POWER, AND REFLECTION

1. L. E. Kinsler, A. R. Frey, A. B. Coppens, and J. V. Sanders, *Fundamentals of Acoustics,* 3rd ed. (New York: John Wiley and Sons, 1982), Chapters 5 and 6. Many basic physical principles of acoustic wave propagation, including the wave equation, energy and power density, impedance, the decibel scale, and reflection.

CHAPTER 4 ACOUSTICAL PROPERTIES OF BIOLOGICAL TISSUES

1. W. Bloom and D. W. Fawcett, *A Textbook of Histology* (Philadelphia: W. B. Saunders Co., 1968). Tissue classification and detailed description of cellular makeup.
2. B. Folkow and E. Neil, *Circulation* (New York: Oxford University Press, 1971). Properties of blood.
3. S. A. Goss, R. L. Johnston, and F. Dunn, "Comprehensive Compilation of Empirical Ultrasonic Properties of Mammalian Tissues," *J. Acoust. Soc. Am.* 64 (Aug. 1978), 423–457. Substantial catalogue of reported tissue data.
4. S. A. Goss, R. L. Johnston, and F. Dunn, "Compilation of Empirical Ultrasonic Properties of Mammalian Tissues, II," *J. Acoust. Soc. Am.* 68 (July 1980), 93–108. Additional data published or discovered since previous paper.
5. D. E. Goldman and T. F. Hueter, "Tabular Data of the Velocity and Absorption of High-Frequency Sound in Mammalian Tissues," *J. Acoust. Soc. Am.* 28 (January 1956), 35–37. Tables of tissue data plus graph of dispersive absorption.
6. R. T. Beyer, *Nonlinear Acoustics,* Naval Ship Systems Command, Department of the Navy (Washington, D.C.; Government Printing Office, 1975), Section 1.6. Discussion of mechanisms of sound absorption and relaxation.
7. C. S. Clay and H. Medwin, *Acoustical Oceanography: Principles and Applications,* (New York: John Wiley and Sons, 1977), 414–419. Absorption and relaxation processes. This book also contains solid technical discussions of transducers, beam patterns, and electronics for ultrasound of more general interest than oceanography alone.

8. K. F. Herzfeld and T. A. Litovitz, *Absorption and Dispersion of Ultrasonic Waves* (New York: Academic Press, 1959). The nature of absorption processes, including relaxation phenomena.

CHAPTER 5 TRANSDUCERS, BEAM PATTERNS, AND RESOLUTION

1. L. E. Kinsler and A. R. Frey, *Fundamentals of Acoustics*, 2nd edition (New York: John Wiley and Sons, 1962), Chapter 12. Covers piezoelectric coefficients and analysis of transducer power, equivalent circuits, and beam patterns.
2. R. T. Beyer and S. V. Letcher, *Physical Ultrasonics* (New York: Academic Press, 1969), Chapter 2. Thorough analytical discussion of piezoelectric transducers and equivalent circuits in addition to other generators of ultrasound.
3. Kynar Piezo Film data sheet from Pennwalt Corp., King of Prussia, PA. Data on PVDF as well as other piezoelectric materials.
4. B. G. Bardsley and D. A. Christensen, "Beam Patterns from Pulsed Ultrasonic Transducers Using Linear System Theory," *J. Acoust. Soc. Am.* (January 1981), 25-30. A general approach to beam patterns for pulse excitation, including results for short-pulsed phased arrays.
5. M. Abramowitz and I. Stegun, *Handbook of Mathematical Functions*, NBS Applied Mathematics Series 55 (Washington, D.C.: Government Printing Office, 1972). Contains a section on Bessel functions, with values and identities.
6. B. D. Steinberg, *Principles of Aperture and Array System Design* (New York: John Wiley and Sons, 1976). Complete analysis of arrays and beam characteristics.
7. P. C. Pederson and D. A. Christensen, "Power Measurement Techniques Applied to Imaging Systems," *Acoustical Holography*, Vol. 6 (New York: Plenum Press), 711-739. Maps of near-field intensity plus theoretical derivation of near-field equations for square transducers.
8. B. Jacobson and J. G. Webster, *Medical and Clinical Engineering* (Englewood Cliffs, N. J.: Prentice-Hall, 1977), 463-477. Discussion of resolution in addition to general clinical ultrasound.

CHAPTER 6 DIAGNOSTIC IMAGING CONFIGURATIONS

1. P. N. T. Wells, "Medical Ultrasonics," *IEEE Spectrum* 21 (December 1984), 44-51. A good overview of the use of ultrasound in medicine.
2. S. N. Hassai and R. Bard, "Evaluating the Eye Through Ultrasonography," *Geriatics* 32 (October 1977), 94-101. Clinical advantages of B-scan in ophthalmology.
3. A. F. Parisi, P. F. Moynihan, B. J. Ray, and D. A. Pieto, "Two-dimensional

Echocardiography," *Journal of Cardiovascular Medicine* (January 1980). Clinical discussion of B-mode and M-mode echocardiography.
4. F. L. Thurstone and O. T. von Ramm, "A New Ultrasound Imaging Technique Employing Two-Dimensional Electronic Beam Steering," *Acoustical Holography,* Vol. 5 (New York: Plenum Press, 1974), 249-259. An early paper giving the concept of a phased-array B-scanner for echocardiography.
5. H. E. Melton, Jr., and F. L. Thurstone, "Annular Array Design and Logarithmic Processing for Ultrasonic Imaging," *Ultrasound in Medicine and Biology* 4 (1978), 1-12. Interesting tradeoffs in the design of a phased-array scanner transducer array.
6. B. G. Bardsley, "A Phased-Array Echocardiographic Imager Using Serial Analog Memory Delay Lines," Ph.D. dissertation, University of Utah, 1981, Chapters 2 and 3. The advantages and tradeoffs in the design of phased arrays.
7. *Hewlett-Packard Journal* (October 1983). This entire issue is devoted to the detailed design of a phased-array imager.
8. J. F. Greenleaf, S. A. Johnson, and A. H. Lent, "Measurement of Spatial Distribution of Refractive Index in Tissues by Ultrasonic Computer Assisted Tomography," *Ultrasound in Medicine and Biology* 3, (1978), 327-339. Gives theory and results (speed of sound and attenuation) for ultrasonic computerized reconstruction.
9. R. K. Mueller, M. Kaveh, and G. Wade, "Reconstructive Tomography and Applications to Ultrasonics," *Proceedings of the IEEE* 67 (April 1979), 567-587. Complete discussion of computerized reconstruction methods and algorithms.
10. R. E. Anderson, "Ultrasonic Holography: A New Tool for Medical Imaging," *Applied Radiology* (January/February 1974). The promises and shortcomings of acoustical holography applied to clinical imaging.
11. C. F. Quate, "The Acoustic Microscope," *Scientific American* (October 1979), 62-70. Good overview of the ultrasound microscope.
12. C. Marwick, "Kidney Stones, A New Era in Therapy for One of Mankind's Most Painful Problems," *Medical World News* (February 14, 1983), 76-93. Compares current treatments, including ultrasound, for kidney stones.
13. D. A. Christensen and C. H. Durney, "Hyperthermia Production for Cancer Therapy: A Review of the Fundamentals and Methods," *Journal of Microwave Power* 16 (1981), 89-105. Compares the advantages and disadvantages of ultrasound as a measure of producing hyperthermia compared to electromagnetics.

CHAPTER 7 DOPPLER AND OTHER ULTRASONIC FLOWMETERS

1. P. N. T. Wells, "Medical Ultrasonics," *IEEE Spectrum* 21 (December 1984), 44-51. Good coverage of contemporary Doppler configurations.
2. L. C. Lynnworth, "Ultrasonic Flowmeters," Chapter 5 in *Physical Acoustics,* W. P. Mason and R. N. Thurston (eds.), Vol. XIV, (New York: Academic

Press, 1979). Very thorough coverage of many different ways of measuring flow by ultrasound, with mainly industrial emphasis.
3. R. D. Waxman, "Doppler Ultrasound," *Radiology Today* (April/May 1980). Covers cw Doppler design and use.
4. J. G. Webster (ed.), *Medical Instrumentation* (Boston: Houghton Mifflin Co., 1978), Section 8.4. CW and pulsed Doppler flowmeters.
5. D. W. Baker, "Pulsed Ultrasonic Doppler Blood-Flow Sensing", *IEEE Trans. on Sonics and Ultrasonics* SU-17 (July 1970), 170–185. An early paper with a thorough discussion of design considerations for pulsed Doppler.
6. T. C. Evans, Jr., and J. C. Taenzer, "Ultrasonic Imaging of Atherosclerosis in Carotid Arteries," *Applied Radiology* (March/April 1979). The clinical value of a duplex scanner (real-time mechanical B-scanner plus multiple range-gated Doppler) for detecting lesions in the carotid arteries.
7. R. G. Burton and R. C. Gorewit, "Ultrasonic Flowmeter Uses Wide-Beam Transit-Time Technique," *Medical Electronics* (April 1984), 68–73. A description of an advanced transit time device for animal studies.

CHAPTER 8 THE SAFETY AND MEASUREMENT OF ULTRASOUND

1. R. C. Thompson, "The Unknowns of Ultrasound," *FDA Consumer*, HHS Publication No. (FDA) 83-8201 (March 1983). Survey, current thinking regarding the safety of diagnostic ultrasound.
2. W. D. Ulrich, "Ultrasound Dosage for Nontherapeutic Use on Human Beings—Extrapolations from a Literature Survey," *IEEE Trans. on Biomedical Engineering* 21 (January 1974) 48–51. Presents a suggested safety guideline based upon hazards documented up to 1974.
3. P. C. Pederson and D. A. Christensen, "Power Measurements Techniques Applied to Imaging Systems," *Acoustical Holography*, Vol. 6 (New York: Plenum Press, 1975), pp. 711–739. Description of the radiation force method for measuring output power, including cautions and experimental results.
4. R. T. Beyer, *Nonlinear Acoustics*, Naval Ship System Command, Department of the Navy (Washington, D.C.: Government Printing Office, 1975), Chapter 8. Thorough discussion of cavitation.

Index

A-mode, 124-129
   applications of, 133-135
Abdominal scanning, 141-143, 148
Absorption, 54, 214, 218
   dispersive, 89
Acoustic holography, 163-164
Acoustic microscope, 169
Acoustic parameters for tissues, 60-64
Air:
   impedance of, 34
   ultrasound characteristics, 64
Alcohol, 19
Angle of reflection, 26
Angle of transmission, 26-29
Angular frequency, 10
Apodization, 104
Attenuation constant, 56
   in tissues, 53-60
   of water, 59
Average power density, 25
Axial resolution, 85-89

B-Mode, 135-158
   applications of, 141-143
   compound, 140-151
   electronic scanners, 146-158
   linear array, 146-149
   manually scanned, 137
   mechanical scanners, 144-146
   phased array, 149-158
   real-time, 143-158
Barium titanate, 76-77, 81
Beam patterns of transducers, 89
Bessel functions, 96
Bibliography, 225
Blood:
   acoustic values of, 61
   ultrasound characteristics, 51-53
Bone:
   acoustic values of, 61
   ultrasound characteristics, 50-51, 64
Blood flowmeter, 175
Boundary conditions, 29-30
Brain, acoustic values of, 61
Breast, acoustic values of, 61
Brillouin diagram, 17
Brillouin equation, 10

C-Mode, 160
Cavitation, 215
Cells, 44-45
Characteristic impedance, 22
Chemical effects, 214
Circular transducer, 92, 96

Classification of diagnostic instruments (table), 124
Compressibility constant, 9
Compression, logarithmic, 89, 132–134, 140
Compressional waves, 6, 30, 71
Computerized tomography, 165
Condensation, 9
Connective tissue, 48–53
  dense, 50
  loose, 50
Conservation of mass, 8
Continuity:
  of pressure, 29, 32
  of velocity, 29, 32

Damage, possible mechanisms of, 214–215
Decibel scale, 37, 62
Demodulation, 132, 134
Dense connective tissue, 50
Density, 8, 63
Depth of focus, 111, 152
Diathermy, 167–168
Directional factor, 97, 101
Dispersion, 17
Dispersion diagram, 16–17
Dispersion equation, 10, 15
Dispersive absorption, 89
Divergence angle, 98, 102
Doppler configurations, 181–183
Doppler flowmeters, 202–203
  analog-output, 190–192
  angles, 183–184
  audible-output, 188–190
  bidirectional, 190–192
  cw, 185–192
    transverse scanners, 194
  duplex scanners, 200–203
  imagers, 192–203
  mixing theory, 186, 191
  pulsed, 195–200
  spectral spread, 184–185
  spectrum-output, 192–193
  volume flowrate, 200, 203, 206
Doppler imagers, 192–203
Doppler principle, 176–179
Doppler shift, 179–181
  equation, 179
  sensitivity vector, 180
Doppler spectrum, 188, 191
Duplex scanners, 200–203

Duty cycle, 216
Dynamic focusing, 154

Echocardiography, 157–158
Echoencephalography, 135
Electronic scanners, B-mode, 146–158
Epithelial tissue, 45–47
Equivalent circuits of transducers, 78–81
Erythrocytes, 52
Eulerian coordinates, 6
Exposure levels, 215–218

f-number, 108
Far field, 90
Fat, acoustic values of, 61
Fermat's principle, 39
Flowmeters:
  doppler, 202–203
  transit time, 203
  vortex, 207–208
Focal length, 106
Focusing with lenses, 104–112
Fourier components, 11
Fourier transform, 129, 131, 185
Fraunhofer approximation, 95
Frequency:
  angular, 10
  temporal, 14

Gray-scale, 139
Group velocity, 17

Heat conduction, 57
Heat production, 167–168
Holography, 163–164
Huygen's principle, 90, 151
Hydrophone, 218
Hyperthermia, 167–168

Impedance, 56
  characteristic, 22
  mismatch, 31, 64, 74
  of air, 34
  of water, 23, 34
  specific acoustic, 21
  tissue, 64
Index of refraction, 162

Initial bang, 125
Intensity, 25
Interfaces, 26, 33

Kidney, acoustic values of, 61

Lateral resolution, 109, 148
Lead zirconium titanate, 77
Lenses:
    focal length, 106
    focusing, 105
Linear array, 112-118, 146-149
Lithotripsy, 167
Liver, acoustic values of, 61
Logarithmic compression, 89, 132-134, 140
Longitudinal waves, 6
Loose connective tissue, 50
Lung, acoustic values of, 61

M-Mode, 158-160
Magnitude of reflection, 29-32
Mechanical scanners, 144-146
Microscope, acoustic, 169
Mode conversion, 30, 51
Momentum, 220
Motional resistance, 79
Multiple reflections, 128
Muscle, acoustic values of, 61
Muscular tissue, 47-48

Near field, 9
Nervous tissue, 48
Newton's force equation, 6
Non-thermal effects, 214, 217-218
Noninvasive imaging, 2
Nonlinear terms, 9
Nyquist sampling criterion, 196

Obstetrical scanning, 141-142, 148, 222
One-dimensional wave equation, 6, 10
Ophthalmology, 135-136

Particle velocity, 6, 21, 23, 29, 31
Peripheral vascular scanning, 194, 202

Phase, 11
Phase fronts, 12
Phase velocity, 13, 15, 17, 23, 56, 63, 127, 203
    of water, 16
Phased array, 149-158
    characteristics and advantages, 156-158
    time delay equations, 154-156
Piezoelectric materials, 69-70
    coefficients, 71, 76-78
    comparison of, 81
Plane waves, 12
Poly(vinylidene fluoride), 77
Power:
    reflection, 34-38
    transmission, 34-38
    total, 24, 36-37
Power density, 24, 36
    average, 25, 216
    measurement techniques, 218-220
    peak, 216
Pressure, 5, 10-11, 29, 31
    continuity of, 29, 32
Propagation constant, 10
Pulse repetition frequency, 128, 146, 196
Pulsed Doppler flowmeters, 195-200
PVDF, 77-78, 81
PZT, 76-77, 81

Quadrature phase detection, 192
Quartz, 74, 77, 81

Radiation force, 219
Radiography, comparison with ultrasound, 2-3
Rays, 14
Rayleigh scattering, 53, 210
Real-time B-mode, 143-158
Rectangular transducer, 100
Reflection:
    angle of, 26
    coefficient, 30-32
    magnitude of, 29-32
    power of, 34-38
    specular, 28
Refraction, 28-29
Relaxation, 59-60
Resolution,:
    axial, 85-89
    lateral, 109, 148

Respiratory monitoring, 207
Rayleigh scattering, 53, 189

Safety, 45, 213
Scattering, Rayleigh, 53, 189
Sectors, 137, 156
Sensitivity vector, 183
Shear waves, 6, 30, 51
Signal conditioning, 129–134
Snell's law, 28, 39
Spatial average of power density, 216
Specific acoustic impedance, 21
Speckle, 218
Specular reflection, 28
Spherical waves, 91
Spot size, 107, 152
Standing waves, 32–35, 72
Surgery with ultrasound, 166–167
Synthetic aperture, 164–165

Temporal average of power density, 216
Temporal frequency, 14
Thermal effects, 214, 216–217
Through-transmission, 160
Time gain control (TGC), 129–132, 134, 140
Time motion, 158
Time of arrival, 125
Tissue impedance, 64
Tissue survey, 43–53
Tissues:
    acoustic parameters for, 60–64
    attenuation in, 53–60
    blood, 51–53
    bone, 50–51
    cells, 44–45
    connective, 48–53
    epithelial, 45–47
    muscular, 47–48
    nervous, 48
    table of acoustic parameters, 61
    viscosity relaxation in, 59–60
Tomographic scan, 139
Total internal reflection, 40
Transducers:
    absorber, 88
    antenna pattern, 98
    beam patterns, 89
    capacitance, 79
    comparison of materials, 81
    continuous wave excitation, 72
    depth of focus, 111–112, 152
    divergence angle, 98, 102–103
    electrical excitation of, 70–72
    equivalent circuits of, 78–81
    f-number, 108
    far-field pattern of:
        circular, 96–100
        rectangular, 100–102
    focused pattern, 104–112, 152
    Fraunhofer region, 103
    frequency response, 73
    Fresnel region, 103
    Gaussian profile, 104
    grating lobes, 114–116, 151
        reduction of, 116–118
    linear array, 112–118, 146–149
    near field to far field transition, 93–94
    near-field pattern (on-axis) of:
        circular, 91–94
    phased array, 149
    piezoelectric coefficients, 76–78
    pulse width, 84
    pulsed excitation, 82
    Q, 73–74, 84, 117
    quartz, 74
    radiated power, 74–76
    resistance, 79
    resonant frequencies, 72
    smoothing profile of, 104
    spot size, 107, 152
Transit time flowmeter, 203
Transmission:
    angle of, 26–29
    power of, 34–38
Transmit/receive (T/R) switch, 125–127
Transverse waves, 6
Traveling waves, 11

Ultrasound, definition of, 5

Velocity:
    continuity of, 29, 32
    group, 17
    particle, 6, 23, 31
    phase, 13, 15, 17, 23, 56, 63, 127
    profiles, 200, 205, 210–211, 212
Viscosity, effects of, 54–56
Viscosity relaxation in tissues, 59–60
Vortex flowmeter, 207

INDEX

Water:
   attenuation of, 59
   impedance of, 23, 34
   phase velocity of, 16
Wave equation:
   derivation of, 5–10
   one-dimensional, 6, 10
   solutions to, 10–11
Wavelength, 12, 14
Wave nature, 11–17
Waves:
   compressional, 6, 71
   longitudinal, 6
   plane, 12
   shear, 6
   spherical, 91
   standing, 32–35
   transverse, 6
   traveling, 11
Width of beam, 102–104

X-radiation, 2, 214

CPSIA information can be obtained at www.ICGtesting.com
Printed in the USA
268216BV00001B/78/A